P9-DBM-379

DATE DUE

Feb 98

MAR 0 5 1998	MAR 0 7 2001	
MAR 1 4 1998	MAR 2 2 2001	
APR 1 3 1998	MAY 1 0 2001	
APR 3 0 1998	DEC 3 1 2002	
	JUN 1 0 2003	
MAY 2 7 1998	JUL 3 0 2003	
	OCT 2 2 2003	
JUN 2 1 1998	DEC 2 7 2003	
JUL 0 6 1998	JAN 3 1 2005	
AUG 1 3 1998	MAR 1 6 2005	
OCT 0 9 1998		
FEB 2 0 1999		
JUN 2 9 1999		
DEC 0 2 1999		
AUG 0 7 2000		

Demco, Inc. 38-293

Yesterday's Toys

Yesterday's Toys

734 Tin and Celluloid Amusements from Days Gone By

By Teruhisa Kitahara

BLACK DOG
& LEVENTHAL
PUBLISHERS

Yesterday's Toys - Celluloid Dolls, Clowns, and Animals:
        Copyright © 1988 by Graphic-sha Publishing Co., Limited
Yesterday's Toys - Planes, Trains, Boats, and Cars:
        Copyright © 1988 by Graphic-sha Publishing Co., Limited
Yesterday's Toys - Robots, Spaceships, and Monsters:
        Copyright © 1988 by Graphic-sha Publishing Co., Limited

This edition published by arrangement with:

 Chronicle Books, San Francisco, CA

Published by:

Black Dog & Leventhal Publishers, Inc.
151 West 19th Street
New York, NY 10011

Distributed by:

Workman Publishing Company
708 Broadway
New York, NY 10003

Printed and bound in Singapore

j i h g f e d c b a

ISBN: 1-884822-95-9

Table of Contents

Yesterday's Toys --

    Celluloid Dolls, Clowns & Animals   13

Yesterday's Toys --

    Planes, Trains, Boats & Cars   117

Yesterday's Toys --

    Robots, Spaceships, & Monsters   221

YESTERDAY'S
TOYS
Celluloid Dolls, Clowns, and Animals

YESTERDAY'S TOYS

Celluloid Dolls, Clowns, and Animals

Teruhisa Kitahara ▪ Photography by Masashi Kudo

THE AGE OF CELLULOID

In the middle of the Meiji era (1868–1912), imported celluloid was worked by hand and used as a substitute for natural materials such as ivory, tortoiseshell, and coral. It was made into ornamental hairpins, eyeglass frames, billiard balls, and other objects. Later, new techniques were developed. The most important of these were:

Pressed Ball Molding: After the celluloid was cut into plate form, it was put in hot water. Once submerged, it was put between semicircular male and female dies, and molded by the heat of the water. The rough edges of the molded hemispheres were then removed, and the two hemispheres bonded together to make a sphere. Products manufactured by this technique included ping-pong balls, self-righting toys, and rattles.

Blow Molding: Celluloid plates were placed between heated male and female dies. Heated compressed air was then injected into the dies to mold the celluloid. The molding process was finished by cooling the dies quickly with water. To make it easier to remove products from the dies, they were coated with liquid soap, which was expensive at that time. This was the forerunner of present blow-molding techniques. Most hollow products—such as Kewpie dolls—were manufactured by this method.

Competing with toys made of traditional materials such as clay, wood, paper, and cloth, celluloid dolls and toys were introduced on a full scale in the late Meiji and early Taisho eras (there are records of celluloid toys being sold in Japan in 1913, but the type of product is unknown). In that age, with the lingering atmosphere of the Edo period, it is easy to imagine people being captivated by this light, versatile, easy-to-color material. The age of celluloid was created overnight. In addition to its beauty and popularity, celluloid had another advantage: camphor, its main ingredient, was cheap, because it was a product of Taiwan. During World War I, Germany, the principal exporter of celluloid toys, halted its production of toys, and this provided Japan with the opportunity to meet world needs. After World War I, there was a sluggish period, but in 1927 and 1928, Japanese production rose to become the highest in the world.

Celluloids were found everywhere in daily life, although their inflammability was a disadvantage: celluloid was used for pen cases, writing boards, subway ticket holders, pails, wash bowls, rulers, bathtub toys, teething rings—the list goes on and on.

However, as the war front in China expanded, the supply of materials for celluloid became short, and with the outbreak of World War II in 1941, production of celluloid dolls and toys stopped completely.

Manufacturers resumed production right after the war. This met the needs of Japanese children hungry for playthings; and supported by a now-unbelievable exchange rate, celluloid exports grew to be the most significant area of total Japanese toy production.

After that, in addition to prewar designs, new products were developed combining special mechanisms with celluloid. This resulted in a string of remarkable products—there were dancing dolls, merry-go-rounds, crawling dolls, and many other toys with new functions. These products quickly helped overcome some of Japan's postwar economic problems. They were actively exported all over the world to obtain foreign currency, each bearing the notification "Made in OCCUPIED Japan."

But soon, celluloid's one defect—its inflammability—became an issue overseas. Non-flammable and flame-resistant celluloids were studied around 1950. In the mid-fifties, the problem was solved by the birth of plastics.

Mr. Kitahara, an enthusiastic, world-famous collector of metal toys, is now publishing this book about the celluloid dolls and toys he has collected. It is very rare to find such a wealth of material on inflammable and breakable celluloid. We respect his passion for preserving these rapidly disappearing toys and applaud his books.

1 11x7x21.5/UNKNOWN/W/1930'S

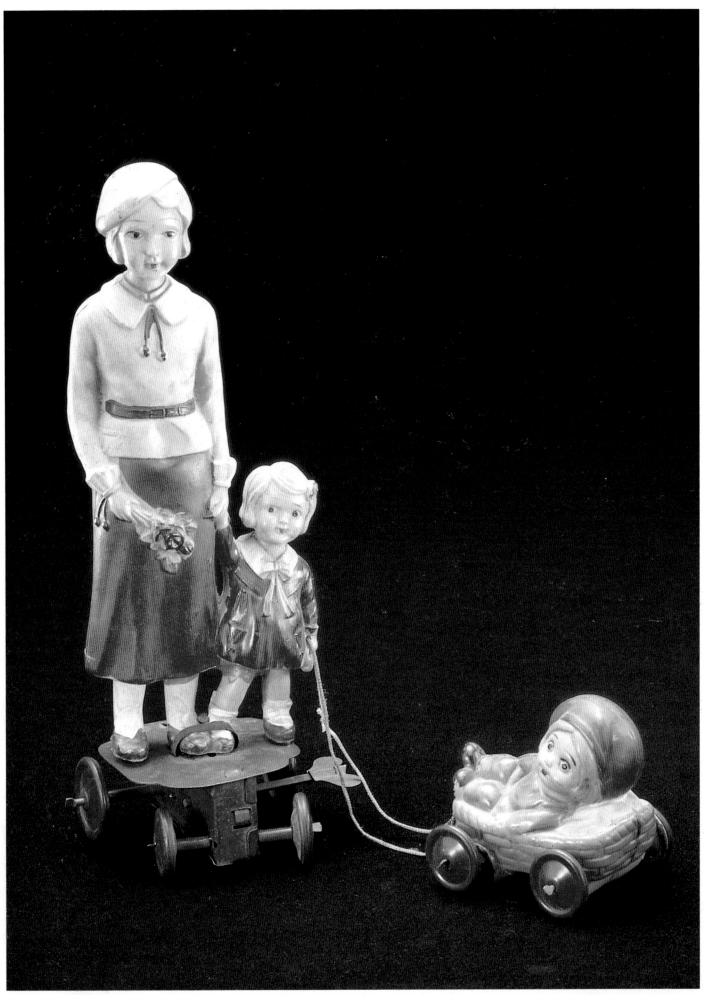

2 8x7x22/O.K. TOY/W/1930'S This toy was made in the 1930s. The fashion trends of the times are apparent in the clothing that the mother and child are wearing.

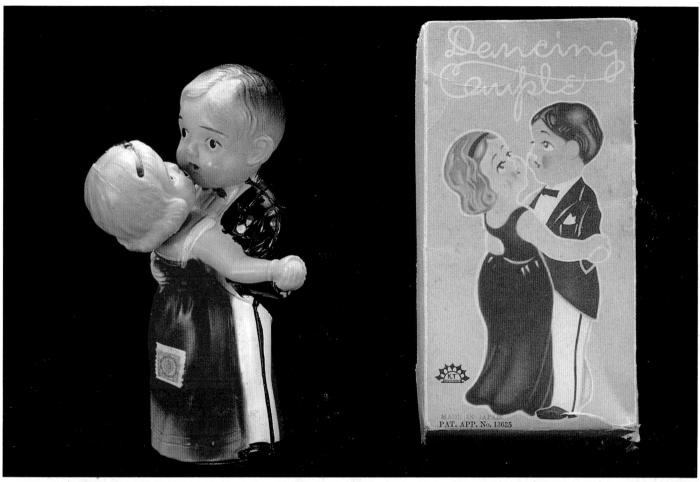

3 5.5x6x15/UNKNOWN/W/1930'S   The way in which these figures turn around—"dancing"—is certainly cute. There have been many types of dancing dolls, but this model was made before the war. Social dancing was very popular then.

4 7x12.5x15/UNKNOWN/W/1930'S   This toy clangs and pedals furiously when wound up. This toy has been designed with great attention to detail, and such fantastic facial expression rarely appears in these toys.

5, 6 7x5.5x17, 5x5x20.5/UNKNOWN/W/1930'S Figure 5's forward motion is created by vibrations coming from within the toy. The same system has been used for the Popeye in Figure 6. Celluloid was chosen for this toy because it is so light. Wimpy can bow.

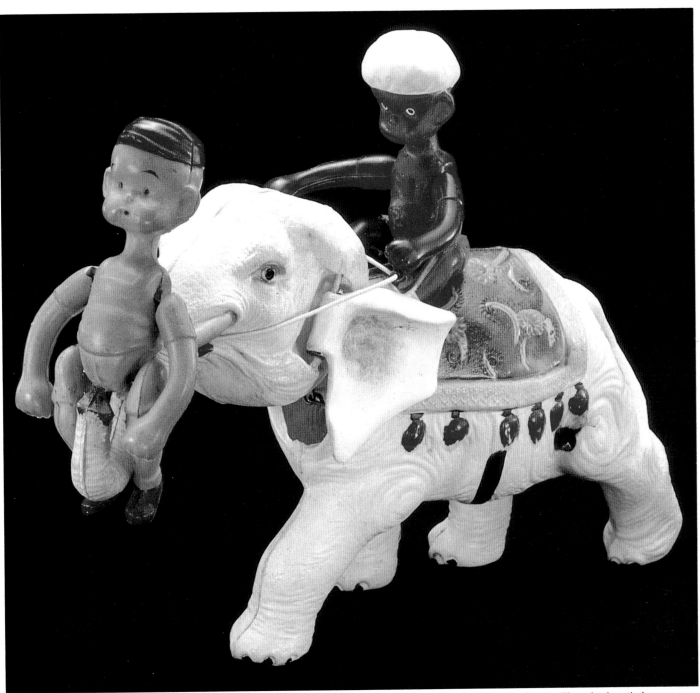

7 6x22x16/UNKNOWN/W/1930'S Henry, a cartoon character, rides up on the trunk of an elephant. The elephant's legs move by vibration, as do its head and ears. The elephant is particularly realistic.

8 6.5x7x14/KURAMOCHI/W/1930'S

Figures 8 and 9 were constructed in the thirties. The lamb moves along beside Mary, nodding its head like a papier-mache tiger.

9 6.5x6.5x16.5/KURAMOCHI/W/1930'S

10 6x6x16/UNKNOWN/W/1940'S   When wound, the umbrella revolves. The popularity of this model continued for a long time.

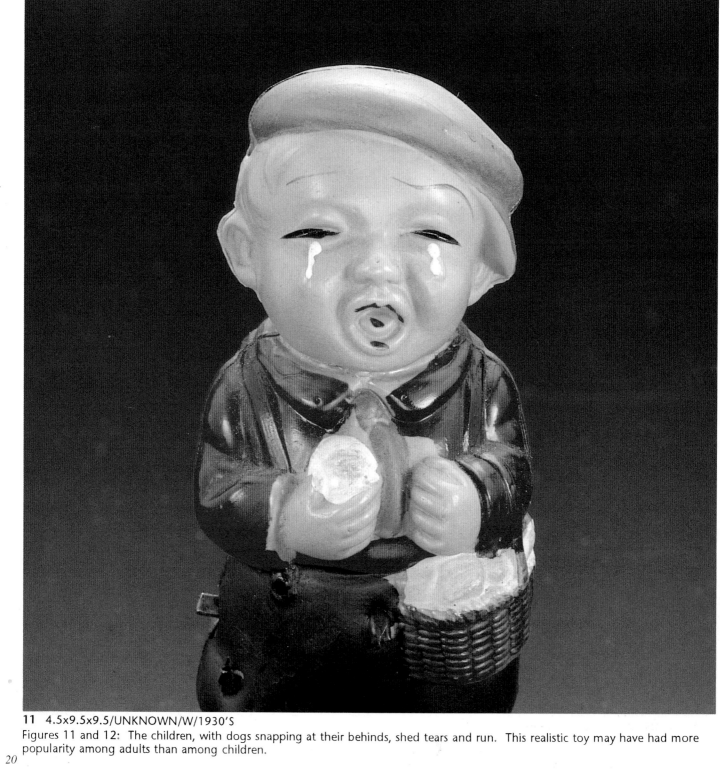

11 4.5x9.5x9.5/UNKNOWN/W/1930'S

Figures 11 and 12: The children, with dogs snapping at their behinds, shed tears and run. This realistic toy may have had more popularity among adults than among children.

12 7x15x14.5/UNKNOWN/W/1930'S

Figures 13 through 23: These trapeze toys were designed by Masayoshi Fujii. The dolls make quick and realistic gymnastic movements. It was a popular toy before the war, and its popularity has continued right up to the present day.

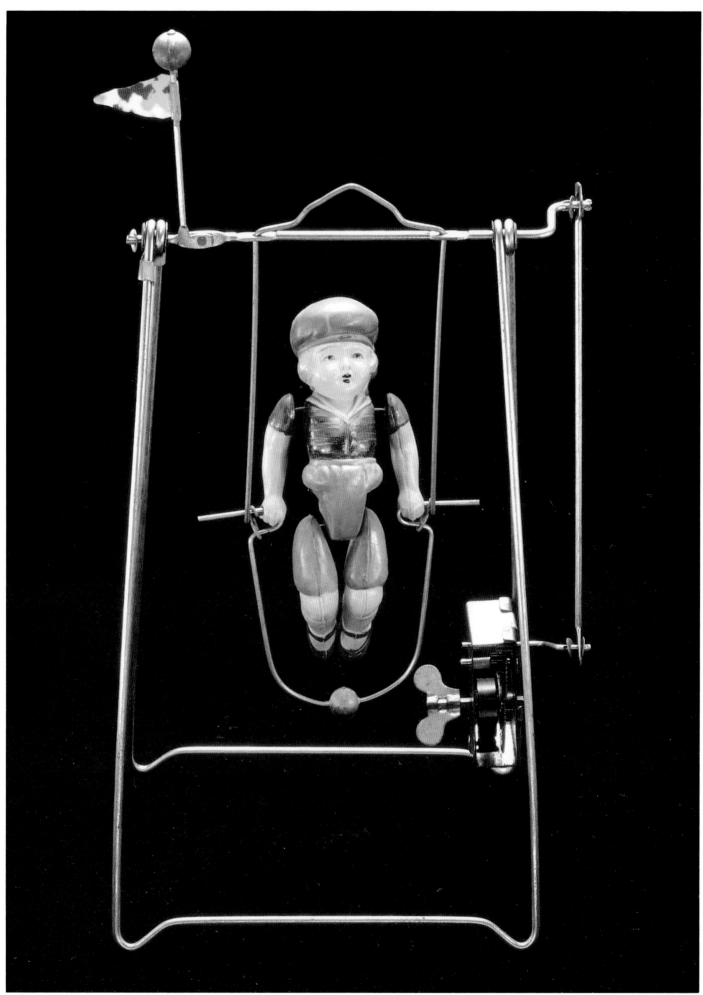

13 13.5x8x23/KURAMOCHI/W/1930'S

17, 18 14.5x20x44,
14.5x23x43/UNKNOWN/W/1930'S

19, 20 10x3x17/UNKNOWN/W/1930'S

14–16 14x6x22/UNKNOWN/1930'S

21 13x3x16/UNKNOWN/W/1940'S

22, 23 14x17x19/UNKNOWN/W/1930'S

24 22.5x11x28/KURAMOCHI/W/1930'S The movement of
the disc underneath the theater makes it look as though the
two cute dolls dressed in clown suits are dancing around on
the stage.

25 16.5x12x25/ALPS/W/1930'S   The deck chair in which the little girl sits rock gently back and forth, and the parasol and duck revolve. A very peaceful scene indeed.

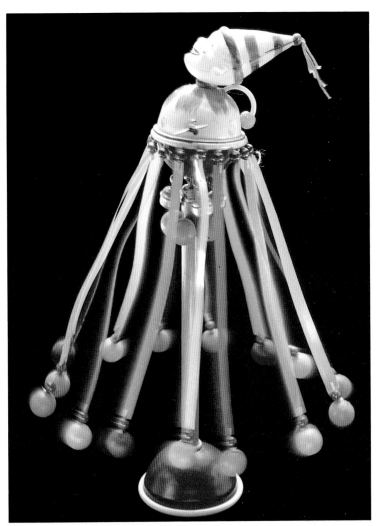

26 8.5x8x33.5/UNKNOWN/W/1930'S

27 24.5x8x33/KURAMOCHI/W/1930'S ·

Figures 26 and 27 : Both of these toys move on a rotary basis. When the clown is spun around the rattles spread out due to centrifugal force. The same principle has been used for the windmill and its bells. These toys were made especially for very small children.

28–30 5.5x5x18, 4x4x13/M.S. TOY/1930'S

Figures 28 through 38: These dolls have no movable parts. All were made before the war. After the war, such realistic faces and clothes disappeared. The parasol in Figure 38 is especially beautifully made.

31–33   5x4x14.5, 4x3x12.5, 4x3.5x12/UNKNOWN/1930'S

34 4.5x3x12.5/UNKNOWN/1930'S

35 7x6.5x20/UNKNOWN/1930'S

36 5x4x15.5/UNKNOWN/1930'S

38 8x5x18/UNKNOWN/1930'S

37 6x4x18.5/UNKNOWN/1930'S

39 9x7x13/MASUDAYA/W/1930'S   This doll powders her face. She seems to be awfully precocious for her age, but the awkwardness of her movements makes her rather sweet.

40 10.5x8x24.5/UNKNOWN/1930'S

41 12x12x20/UNKNOWN/1930'S

42 13x6x13/UNKNOWN/1930'S This doll has no movable parts. The combination of a baby and cushion is quite unusual. Presumably, it was used as a wall decoration.

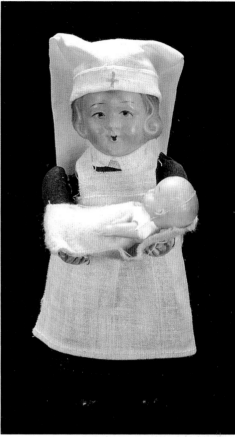

43 7x15x13/UNKNOWN/W/1930'S The doll follows the baby carriage. The shapes and details of both the carriage and the doll are exceptionally well done.

44 5x9x13.5/UNKNOWN/W/1930'S The nurse/doll tries to keep the baby amused by moving her arms up and down.

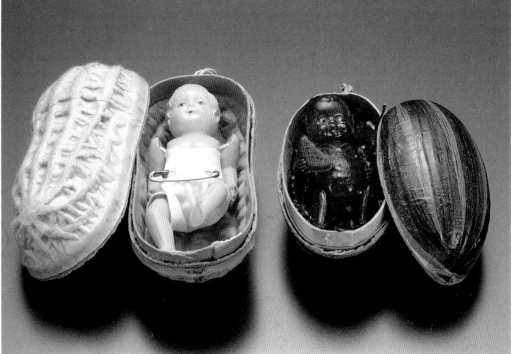

45, 46 5x6x10, 4.5x5x8/MARUGANE/1930'S These celluloids are Christmas tree decorations. The peanut and watermelon open to reveal babies inside.

47 5x5x12.5/UNKNOWN/1930'S

48 12x4.5x17/UNKNOWN/W/1930'S

49 22x20x19/UNKNOWN/1930'S

50 6x4.5x15.5/UNKNOWN/1930'S

51 17x7x17/KURAMOCHI/W/1930'S

52 19x7x21.5/KURAMOCHI/W/1930'S
Every piece of fruit in the wheelbarrow was
modeled and painted by hand. It must have
taken hours. Owing to the fact that each
part was hand painted, the colors differ from
fruit to fruit.

53 15x6x15/UNKNOWN/W/1930'S   The clown makes a comical face while he sits on top of the rattle that is being pulled by the horse. When the rattle revolves, the hanging ornaments spread out and the music starts.

54 10x8.5x36/KURAMOCHI/W/1930'S   This figure
balances on his forehead a rotating baton atop a
wooden pedestal. The act of balancing seems almost
real in this contorted figure.

55 14x14x31/KURAMOCHI/W/1930'S   When the dance in the
center revolves, centrifugal force sends the attached balls
spreading out around her. It is a truly beautiful sight. Such an
attractive effect could only be produced in the medium of
celluloid.

56 6x20x18/UNKNOWN/W/1930'S   The horse moves forward by leaping up and down due to the fact that its hooves are attached to wheels.

57 5x12x21/KURAMOCHI/W/1930'S   As the horse advances via vibrations, the box the clown balances on his forehead revolves. The horse is made of tinplate, the clown of celluloid, and the box of paper.

Figures 58 through 62: These drummers were created before the second World War and display various mechanical methods. Figure 60 drums by means of vibrations that run through its body; the others hit their drums with independently moving arms.

58 13x12x30/KURAMOCHI/W/1930'S

59 9x8x30/UNKNOWN/W/1930'S

60–62 5.5x10x19.5, 5.5x6x15.5/UNKNOWN/W/1930'S

63 7x21x23/MASUDAYA/W/1930'S   The baby lying on its back shakes its head back and forth while revolving the drum by the alternate use of both its arms and its legs.   The stitches of the child's sweater are rendered in great detail.

64, 65 The model for this dog playing the drum is taken from a television cartoon character. It nods its head while playing the drum.

66 17x9x20/UNKNOWN/W/1930'S This figure has moving hands and head. The box underneath contains a music-box device.

68 7.5x7x24/KURAMOCHI/W/1930'S

Figures 67 and 68: Celluloid toys made before the war have exceptionally realistic facial features—Figure 67 has more expression than rabbits generally have. Figure 68 produces music in the same way that Figure 66 does. It also moves its arms and ears.

67 4x3.5x12.5/UNKNOWN/1930'S

69 15x8x26/UNKNOWN/1930'S No part of this figure is movable. It is quite an unusual color for a bear. It's not such a loud color; really, it's a very pleasant shade of green. It also has a lovely face and fur.

70 5.5x5.5x15.5/UNKNOWN/1930'S

71, 72 15x8.5x14/UNKNOWN/1930'S   These little angels dance around. They are operated by a bellows that is connected to a wire that projects from the bottom of the box.

73 15x11x16/KURAMOCHI/W/1930'S

74 12x4x8.5/UNKNOWN/1930'S

75 7.5x14x10/UNKNOWN/1930'S

Figures 76 through 79: These figurines have no movable parts. Soldiers and boy scouts such as these were made in Japan for export use. It is interesting to note that these patriotic toys were made in Japan.

76 5x4.5x15.5/MARUGANE/1930'S

77 3x9x12.5/UNKNOWN/1930'S

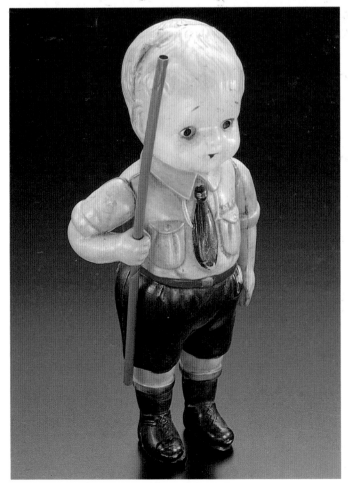

78 6.5x5x15/MARUGANE/1930'S

79 5.5x4x14.5/UNKNOWN/1930'S

80   14x9x30/UNKNOWN/1930'S

81   6.5x4.5x17/KURAMOCHI/W/1930'S

Figure 81: This Gentleman Ventriloquist was the prototype for Edgar Bergen's Charlie McCarthy. It moves by means of vibrations while opening and closing its mouth.

82 9x5x14.5/UNKNOWN/1930'S

Figures 82 and 83: Figure 82 does not move. The authoritative position of its hand is maintained by means of sand inside, which acts as ballast. Figure 83 is a wind-up. It moves its hand in accordance with the traffic signal next to it.

83 16x8x32/KURAMOCHI/W/1930 S

85 3.5x18x14/K.T. TOY/W/1930'S

84 5.5x11x20/UNKNOWN/W/1930'S This peculiar figure moves forward by vibration while raising and lowering his chicken, which seems to be struggling to free itself.

86, 87 2.5x2.5x9.5, 5x5x19.5/UNKNOWN/ 1930'S

88 4x16x13.5/UNKNOWN/1930'S

89 8.5x7x14.5/UNKNOWN/1930'S

90 7.5x7x16/UNKNOWN/1930'S

Figures 88 through 90: These dolls give an indication of the style of the outfits that prewar athletes used to wear.

91 9x9x15.5/UNKNOWN/1930'S

92 9x9x12.5/UNKNOWN/1930'S

Figures 91 and 92: These celluloid dolls are elaborately clothed in painstakingly worked dresses.

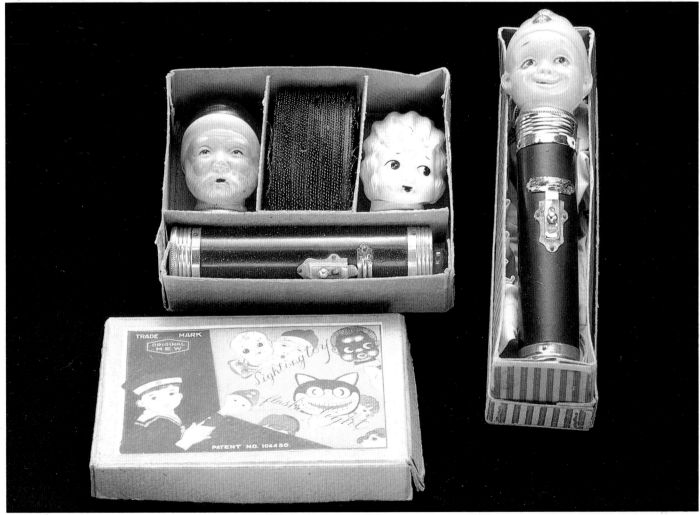

93, 94 14.5x10.5x5.5, 5x19.5x4.5/UNKNOWN/B/1930'S   These celluloid dolls' heads were to be attached to the head of a flashlight. This was not really a practical idea, considering celluloid's inflammability, but the heads could be changed, and the package came with a bulb-cleaning brush.

95, 96 8.5x12.5x16, 7x14.5x17/UNKNOWN/W/1930'S

97 8x6x22.5/UNKNOWN/1930'S

98 9x6x30/UNKNOWN/1930'S

99 7.5x6.5x22/UNKNOWN/1930'S

100 11x7x21.5/UNKNOWN/W/1930'S This model rocks its body from right to left while playing the violin. The head is made of celluloid, but the rest is tinplate.

101 8x6x22.5/UNKNOWN/1930'S

102 4x4x15.5/UNKNOWN/1930'S

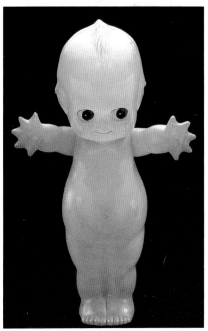

103 10x8x34/UNKNOWN/1930'S

104 8x4.5x2.5/UNKNOWN/1930'S   These toys were quite common at fairs after the war, but this particular model was made before the war. The camphor is inserted in the stern of the boat, and it is then set in the water. As the camphor dissolves in the water, it propels the celluloid boat forward.

105 19x4.5x8/UNKNOWN//W/1930'S   This model boat actually moves along the ground on wheels. The striking difference of scale between the fishermen and the boat is fascinating.

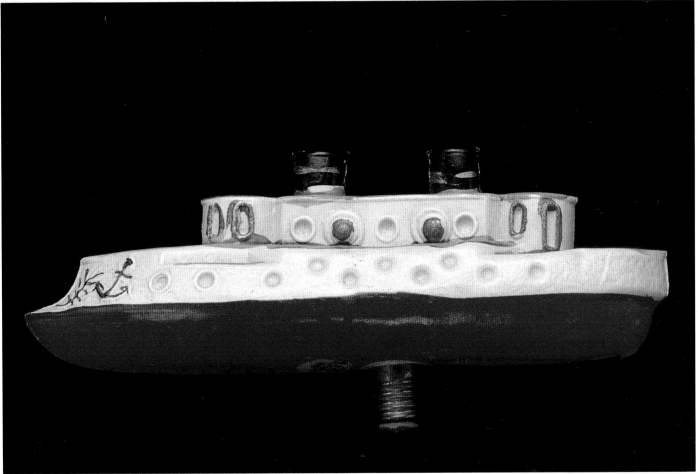

106 3.5x12.5x5/UNKNOWN/1930'S   This figure is a Christmas ornament. It can be lit by a lamp which is placed inside. When it glows in the dark, one can almost imagine the happy cries of the passengers and the lapping of the waves against the hull.

MADE IN JAPAN

107, 108 4.5x18x13/UNKNOWN/W/1930'S When they are placed on the floor fully wound up, these horses will move forward with vigorous jerks of their legs. The legs will not move unless they are brought into contact with the floor.

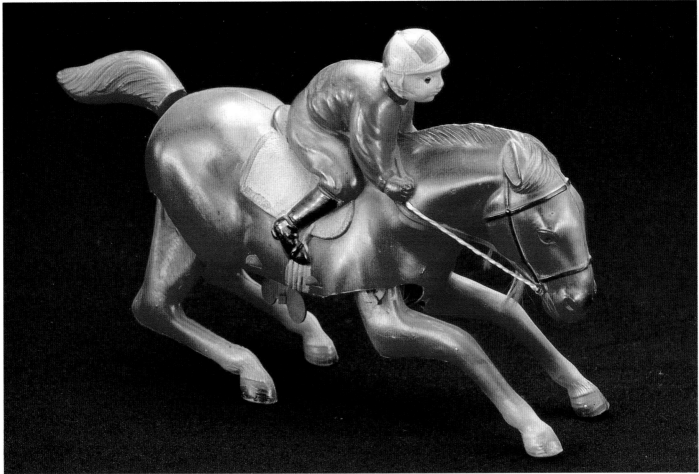

109 4.5x18x13/UNKNOWN/W/1930'S Going by the blue ribbon attached, it is a safe assumption that this particular model was destined for export to Great Britain.

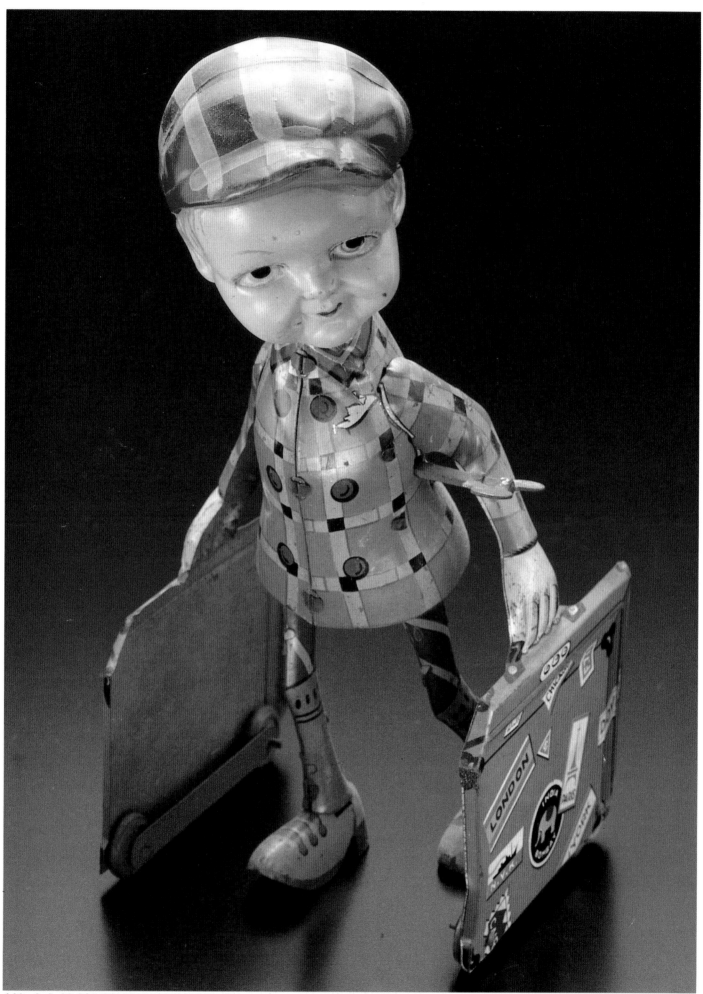

110 9x8x19.5/KURAMOCHI/W/1930'S   This traveler appears to be pulling along two suitcases. In fact, the suitcases are supporting the figure as it moves along step by step.

111 7x6x17/KURAMOCHI/W/1930'S   This toy was used as an advertisement. Although it is quite ordinary in its movements, its facial expression and tartan check coat stick in people's minds.

112 7.5x6x20/KURAMOCHI/W/1930'S   The head of this clown shakes up and down and around. His pathetic expression seems very realistic.

113 14x10x30/UNKNOWN/W/1930'S   This figure lurches and twists its body around in imitation of a drunk. It holds a glass in its left hand and a bottle of beer in its right. Its legs are made of wood, while the rest of its body is celluloid.

114 11x12x24/MASUDAYA/W/1930'S This toy soldier was made during the war; he moves his head from right to left while raising and lowering his rifle. The wartime milieu is apparent from the picture on the box.

115 8x10x26/UNKNOWN/1930'S

117, 118 5.5x4.5x14.5/UNKNOWN/1930's

119 9.5x6.5x24/UNKNOWN/1930'**S**

116 4x3.5x12/UNKNOWN/W/1930'S
This figure move forward on the wheels under its feet, stops to turn around, and then advances again. These toys were made for distribution within Japan.

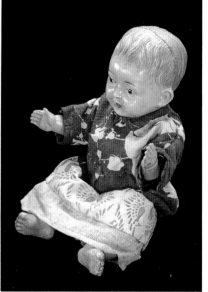

121 9x6x18/UNKNOWN/1930'S

120 4.5x4x10/UNKNOWN/1930'S   This little figure, a cartoon character, does not move.

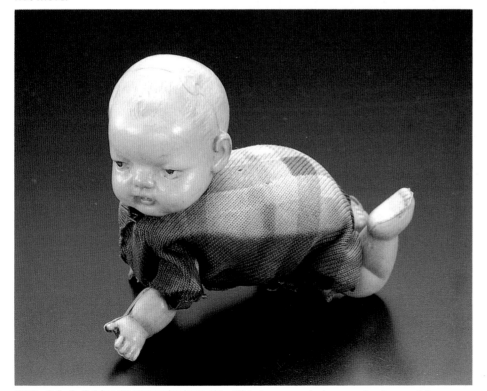

122 9.5x5x7.5/UNKNOWN/W/1930'S   This Japanese baby crawls along while shaking its head from side to side. This particular model was made before the war, but the concept continued in production afterwards.

123 5x5x17/UNKNOWN/1930'S

124 8.5x4.5x14.5/UNKNOWN/1930'S   This figurine has no movable parts. One of the figures carries a rifle and a trumpet. It was obviously made during the war.

125 9x6x20/UNKNOWN/
1930'S

126 6.5x5x14.5/UNKNOWN/1930'S

Figures 126 through 133: The face of Figure 128 moves from left to right by means of elastic and a pendulum. Apart from this, none of the figures have movable parts. All of these Santa Clauses were made before the war.

127 6.5x6x17/UNKNOWN/1930'S

128 7x6x17/UNKNOWN/1930'S

129, 130 6x5x15/UNKNOWN/1930'S, 1940'S

131 5.5x30x7/UNKNOWN/1930'S

132, 133 4x8x9, 5x9.5x10/UNKNOWN/1930'S

134 7x11x11.5/ALSPS/W/1940'S

135 4.5x12x13/UNKNOWN/W/1940'S

136 9.5x3x16/UNKNOWN/1940'S

137, 138 6x5x15, 4x3.5x9.5/UNKNOWN/W/1940'S

Figures 139 through 144: All of these toys were made in occupied Japan. During that time, all Japanese manufacturers were obliged to print "Made in Occupied Japan" on each product. This obligation continued from 1945 to 1950. None of these figures have movable parts.

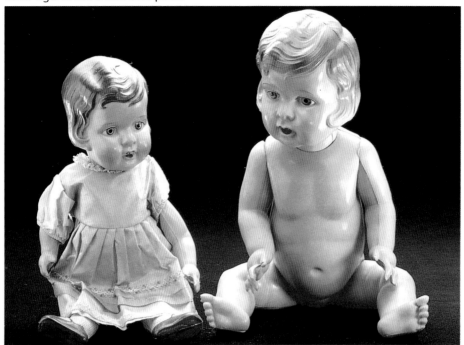

139, 140 12x8x23, 13x8x29/SEKIGUCHI/1940'S

141, 142 7x6x31, 7x5x29/UNKNOWN/1940'S

143 4.5x2x10.5/ROYAL/1940'S

144 6.5x4x16/UNKNOWN/1940'S

145 6x9x11/UNKNOWN/W/1940'S

Figures 145 through 157: All of these toys were made in occupied Japan, and they are all wind-ups. Figures 155 through 157 were Easter toys and were made for export.

146 5x5x11/UNKNOWN/1940'S

147, 148 9x5x12/UNKNOWN/W/1940'S

149 4.5x16x11/UNKNOWN/W/1940'S

150, 151 4x7x10, 4x9x12/UNKNOWN/W/1940'S

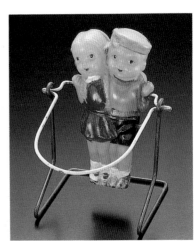

152 7x6.5x12.5/UNKNOWN/ W/1940'S

153 6x6x13/UNKNOWN/W/1940'S

154 11.5x4.5x6/UNKNOWN/W/1940'S

155 5x18x9/UNKNOWN/W/1940'S

156, 157 4x21x10/UNKNOWN/W/1940'S

158 10.5x7x12/KOKYU SHOKAI/W/1940'S   Made in the 1950s, this elephant jumping rope is rather comical. The realistic facial features include such details as the wrinkles on the elephant's nose. The patterns on its clothes are also very cute.

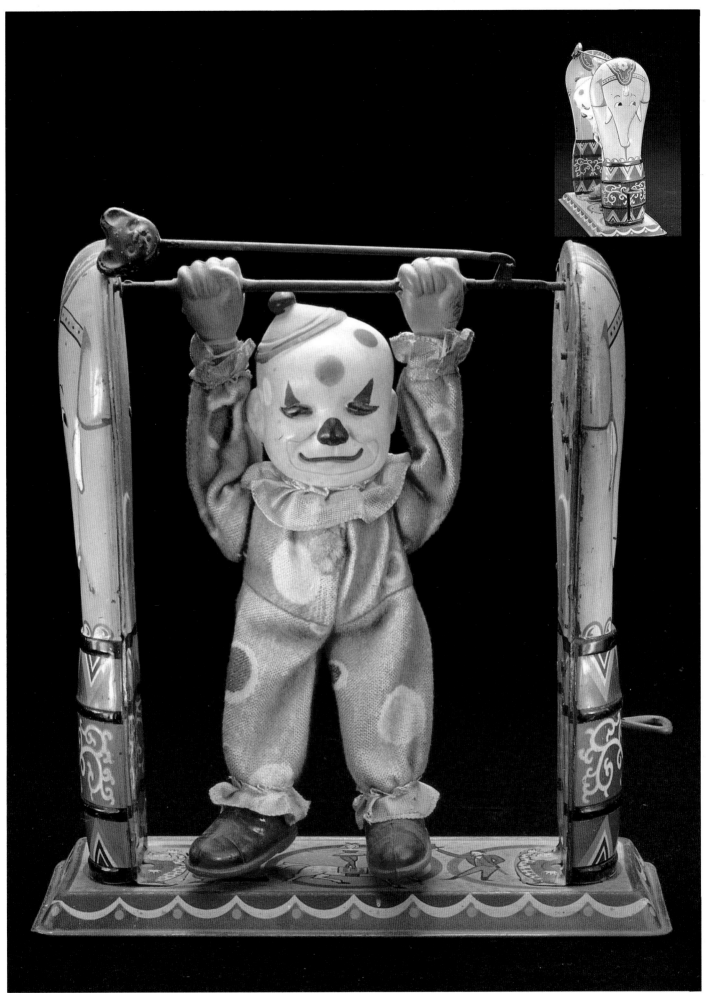

159 18.5x8.5x19/MASUDAYA/W/1950'S   This swinging clown was made in the 1950s.   The posts of his sturdy-looking trapeze are decorated with elephants, but, nonetheless the clown has a melancholy expression on his face.

Figures 160 through 168: These little dolls were meant to be bath toys. There were many such toys produced, but unfortunately, celluloid melts in hot water, so these toys have become quite rare.

160 6.5x16x6/UNKNOWN/1950'S

161 5.5x19.5x6/UNKNOWN/1950'S

162 5x14x6/UNKNOWN/1950'S

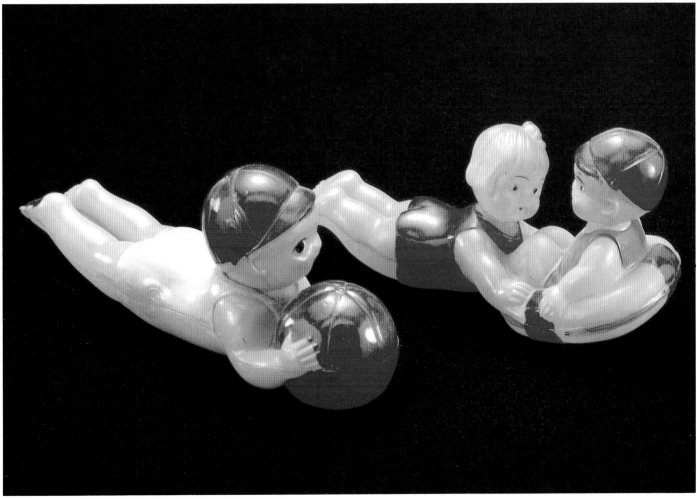

163, 164 5x12x6.5/UNKNOWN/1950'S

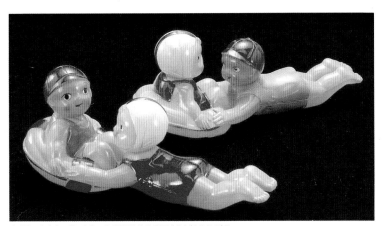

165, 166 5x12x6.5/UNKNOWN/1950'S

167, 168 5.5x18x7.5, 3.5x14x6/UNKNOWN/1950'S

Figure 170: A glass tube containing realistically colored liquid is placed on the back of each of these figures. The warmth generated by a hand holding this tube causes air pressure which ejects the liquid out the front of the tube. The result is a slightly more prurient version of the baby doll who wets.

169, 170   3.5x3.5x8/UNKNOWN/1950'S

171   3x1.5x7.5/SEKIGUCHI/1950'S

Figure 172 and 173: Both of these figures are Halloween toys. The witch's face lights up for a particularly creepy effect.

172 6x6x11/UNKNOWN/B/1950'S

173 5x9x11/UNKNOWN/W/1950'S

Figures 174 through 176: Figures 174 and 175 were produced in the 1950s and do not move. Figure 176 is a creation of the 1930s and nods its head backwards and forwards by the means of elastic and a pendulum.

174, 175 5x4.5x15/UNKNOWN/1950'S

176 6.5x6.5x17/UNKNOWN/1930'S

177 5x9x11/LINE MAR/W/1950'S

178 8.5x2x6.5/UNKNOWN/1950'S   This celluloid doll, representing the legendary Japanese character Momotaro, bursts from a peach made of wood. When closed, an adhesive holds the sections together for approximately ten seconds before the hero pushes his way out again.

Figures 179 through 188: These celluloid Kewpie dolls were made after World War II. They are brighter and more appealing than the prewar models.

179 UNKNOWN/1950'S

180-182 2x7x6/UNKNOWN/1950'S

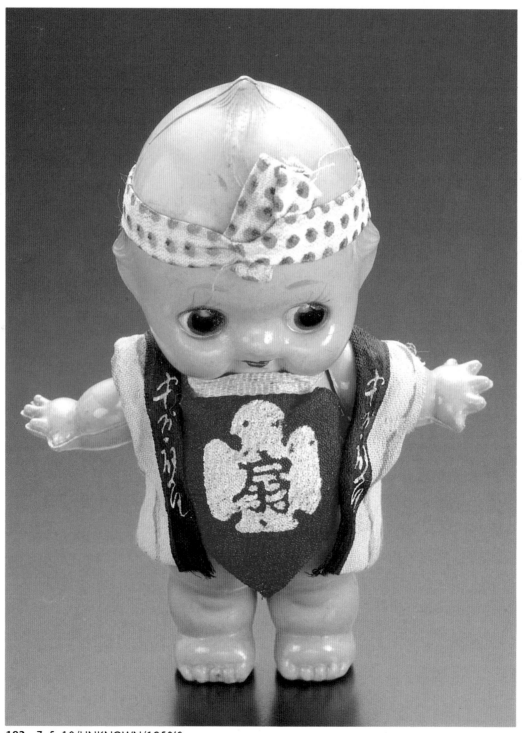

183 7x5x10/UNKNOWN/1950'S

184, 185 7.5x5.5x17, 6x5x13/UNKNOWN/1950'S

186–188 7x6x25, 6x5x19.5, 4x3.5x15/UNKNOWN/1950'S

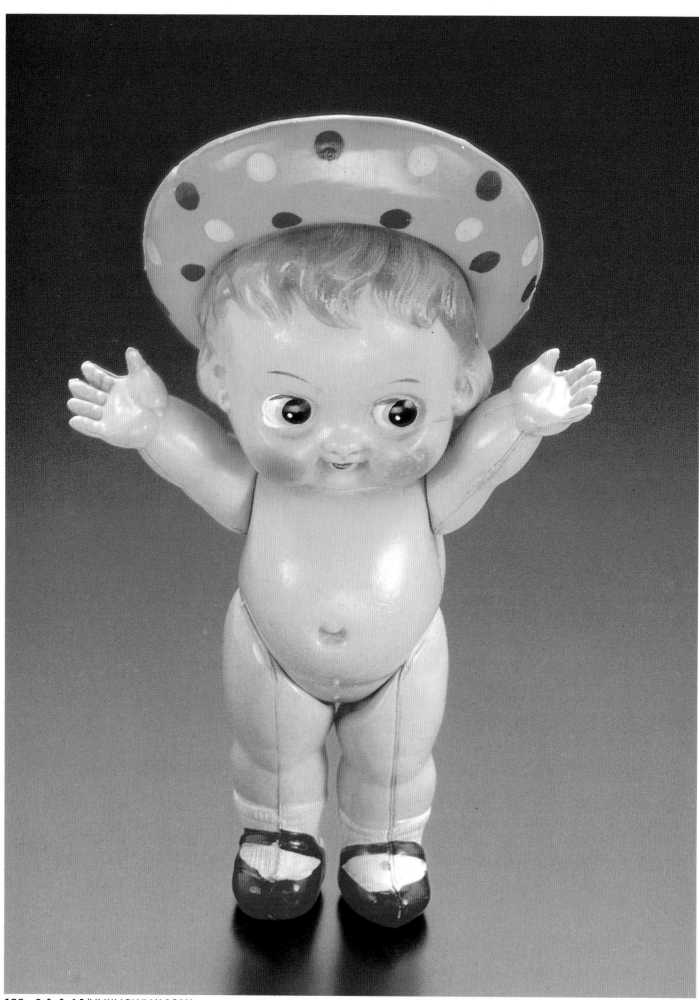

189 9.5x5x15/UNKNOWN/1950'S

190   7.5x6x17/UNKNOWN/1950'S

191   8.5x6.5x20/UNKNOWN/1950'S

192   12x11x25/SEKIGUCHI/1950'S

193, 194   4x3x13.5, 6x3.5/17/UNKNOWN/1950'S

195 14x8x28/UNKNOWN/
1950'S

196 6.5x6.5x11.5/
UNKNOWN/1950'S

197–199 6x5x26, 6x5.5x19, 7x6x30/UNKNOWN/1950'S

200 7.5x7.5x17/UNKNOWN/1950'S

201–202 5x5x17, 6x5x20/UNKNOWN/1950'S

203–207 2.5x2.5x7/UNKNOWN/1950'S

99

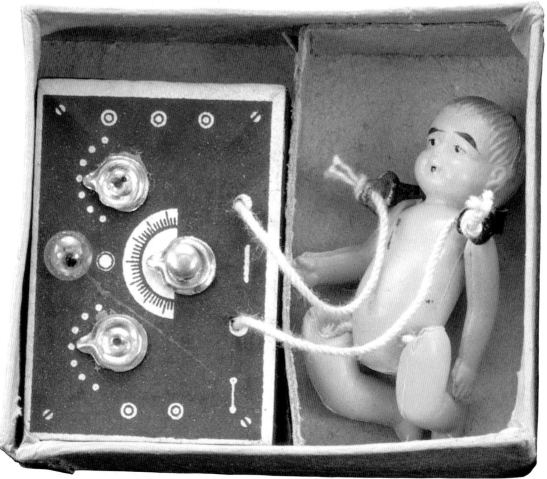

208 8x6.5x3/MARUGANE/1950'S   Dolly's Radio, featuring a baby doll with headphones, has very detailed knobs and dials.

209 7x10x10/UNKNOWN/1950'S

210, 211 11x9x9, 9x7x6.5/UNKNOWN/1950'S

212, 213 6x3.5x13, 5x7x6/UNKNOWN/1950'S

214 22.5x11x28/UNKNOWN/W/1950'S

215 4x3x15/UNKNOWN/1950'S

216 4x3x12.5/
SEKIGUCHI/1950'S

217, 218 8x4.5x19, 7x4x19/MASUDAYA/W/1950'S   Figure 217 is a wind-up. The hand holding the pistol moves up and down while the cowboy's head moves from side to side. Figure 218 has no movable parts.

219 6x10x14/UNKNOWN/F/1950'S The movement of the rocking chair in which the clown is seated slowly pulls the dog towards it, giving the impression of mysterious forces.

220, 221 3x3x6.5, 4x3x6/UNKNOWN/1950'S

222 18x6.5x13/UNKNOWN/W/1950'S The duck quacks by means of a bellows system. The movement of a child chasing a bird is well expressed by the spring which connects the child to the bird.

223–226   3.5x2.5x6/UNKNOWN/1950'S

227   1x1.5x2/UNKNOWN/1950'S

228 3.5x3.5x10/UNKNOWN/1950'S

229 1.5x2x2/UNKNOWN/1950'S

YESTERDAY'S
TOYS
Planes, Trains, Boats, and Cars

YESTERDAY'S
TOYS
Planes, Trains, Boats, and Cars

THE HISTORY OF TIN TOYS

Many people feel nostalgic about tin toys. During their long history, they have earned a place in our hearts. Before about 1872, empty tin cans, mainly petroleum cans, were used to produce simple toys. After that, many tinplate toys were imported—in fact, the Japanese word for tinplate, *buruki,* comes from a Dutch word, *blik*. Imported toys showing modern inventions such as locomotives and ships, gave considerable stimulation to domestic manufacturers. But in those early stages, they did not make copies of those toys; they simply studied them as samples. For a decade, Japanese toys continued to be traditional and provincial—rattles, rickshaws, and little animals.

In 1882, a metal toy with iron wires for spring power appeared. A second important development in this period was the creation in Japan of the flywheel toy, which would be the model for friction toys in later years. After the Sino-Japanese war, the Japanese toy industry was revivified by the victory. There was an instant demand for war toys such as trumpets, sabers, and guns. The popularity of the war was reflected even in toys.

Tinplate toys began to be satisfactory in quantity and quality around 1903 and 1904, at about the time of the Russo-Japanese War. Tinplate-printing machines were imported and press machines were introduced into Japan. Meanwhile, toys with springs began to be produced in huge numbers, stimulated by an elaborate German exhibition of action toys shown throughout the country in this period. The springs in the German toys were made of steel, far superior in performance to the wire and brass springs used in Japan until this time. The introduction of these steel springs contributed greatly to the development of tinplate toys.

In the Taisho era, exports of new battery-operated vehicle toys increased dramatically, and Japanese toy manufacturers enjoyed a new prosperity. This continued through the early years of the Showa era (1925–1989). Nevertheless, in 1938, the production of tinplate toys for domestic use was prohibited as the country conserved resources for the upcoming war. The ensuing sluggishness of the toy industry continued until after World War II. Production and exportation resumed during the Occupation. In 1948, friction toys were created in Japan, and locomotives, fire engines, and cars became the star export toys. Around 1955, electric toys replaced friction toys as the leaders, and in 1963, tinplate toys accounted for 60% of Japan's total toy exports.

Tinplate toys, deeply influenced by historical eras and events, have grown into the toys of today.

1 20x47x18/YONEZAWA/F/1950'S   This 1950s friction toy is a copy of a famous racing car.

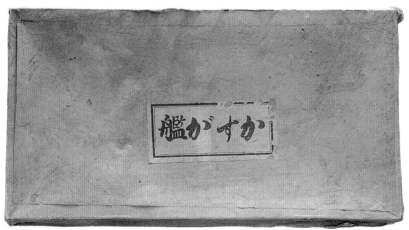

3 3x9x6/UNKOWN/1910'S   This push-along toy is small enough to fit in the palm of one hand This is one of the early Japanese-made penny toys.   The concept comes from German penny toys.

4 2x6x5/UNKNOWN/1910'S   This tiny toy is also a penny toy, but it has extensive detail, particularly on its mast.

5 8x16x9/UNKNOWN/W/1900'S   This car and driver, made during the early part of the century, were painted in red and green in great detail with the use of an airbrush.

6 6x31x9/UNKNOWN/F/1890'S   The train moves, and simultaneously, the string that keeps the wheels moving retracts. This is the original use of the flywheel mechanism.

7 16x6.5x22/KURAMOCHI/W/1920'S   These two seesaws turn around in the same way that a ferris wheel at a fair turns.

8 6x13x9/UNKNOWN/W/1920'S   This truck was made in the 1920s, as we can deduce from the fashionable garb of the riders.

9 8x18x3/UNKNOWN/1900'S These rattles were made around the turn of the century and feature typical Meiji-era scenes. The handle also functions as a flute.

10 4x16x8/UNKNOWN/1930'S This wheeled fish has no mechanical features. It was probably used ornament on Children's Day.

11, 12 9.5x8x9.5, 7x6x12/UNKNOWN/1930'S

13 5x12x11/M.K. TOY/1930'S

14, 15 19x9x9.5/UNKNOWN/1930'S

16 18x15.5x3/UNKNOWN/1930'S   These early tin soldires were inspired by the approaching war. They were very popular in the thirties.

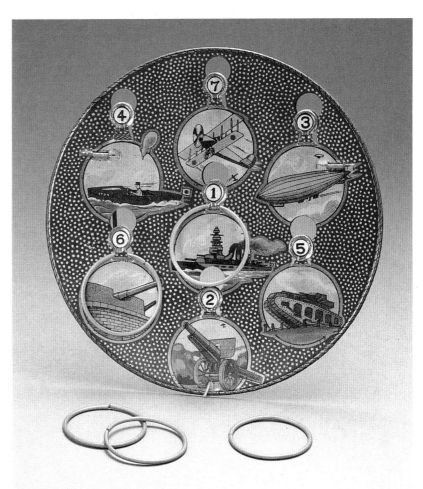

17 17x17x1/Y.C. TOY/1930'S This tin disk is simply a variation on a horseshoe game. The pictures on the disk are of tanks and airplanes, which like Figure 16, reflect the militaristic mood of the thirties.

18,19 11x8x15, 13x7x18/UNKNOWN/1920'S When the handles on these telephones are turned, the bell rings. Figure 18 is made of wood and Figure 19 is made of tinplate.

20 10x36x8/MASUDAYA/W/1930'S   This rather outsized traffic cop directs diminutive trains and cars in a brightly colored city. The beautiful label-studded package features the same gigantic traffic cop.

21 25x13x12/MASUDAYA/W/1930'S This is a tinplate version of the Shanghai incident. The Japanese troops, surrounded by the Chinese army, await an airlift of Japanese troops.

22 5x55x7/UNKNOWN/W/1930'S This army train, replete with cannons and ammunition, is obviously a toy of the war.

23 6x17x7/M.K.TOY/1930'S This toy was made during the period when the Olympics were supposed to be held in Japan for the first time. The five rings, depicting the five continents, are printed on the tank.

24 5x30x6/K.K. TOY/W/1920'S
The first time a zeppelin came to Japan, the sight of the huge airship surprised many people, and it was immediately produced as a toy.

This model hangs by a thread and revolves by means of a screw on the stern of the craft.

25 6x19x15/UNKNOWN/W/1920'S   This moves backwards and forwards by means of a winder. The soldier's hands, which hold the sword and rifle, move up and down with the vibrations.

26 13x23.5x12/MASUDAYA/W/1930'S   This tank advances a certain distance, and it comes to a stop, the lid opens and a soldier pops up. He fires a cannon, drops back inside, and the lid slams shut again.

27 44x36x13/NOMURA/W/1930'S   This is quite a large airplane. While the propeller turns, the plane moves forward. The name of the plane seems to indicate that this toy was copied from an actual war plane.

28 23x22x26/UNKNOWN/W/1930'S   Three airplanes belonging to Japan, the United States, and Great Britain fly around the tower. The facial expression of each pilot represents his country.

29 7x39x13/H.YAMADA/W/1930'S   The rotation of the propeller causes the boat to surge forward, but its direction can be changed by turning the steering wheel.

30 8.5x48x15/UNKNOWN/1930'S

31 8x43x12/UNKNOWN/W/1930'S   This very peculiar model moves forward, via wheels, on land. It manages to give the air of a ship ploughing its way through heavy waves due to the fact that the axle of the wheels is slightly off-center.

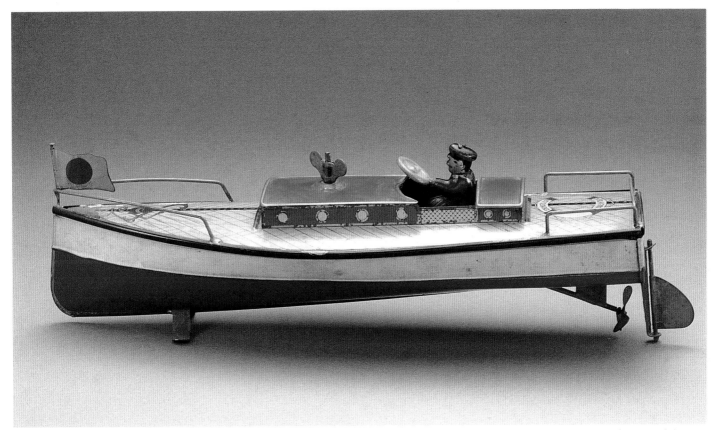

32 8.5/31/9/ENOMOTO/W/1930'S   In its day this toy was very expensive, due to its elaborate detailing. Again, this model can move through the water by means of its propeller.

33 16x6x9/UNKNOWN/W/1920'S   This train moves along a circular track with a diameter of 40cm. The car is a realistic rendition of train cars in the 1920s.

34, 35 2x8x5/UNKNOWN/1920'S   These push-along toys have been painted to resemble the trains that were decorated with flowers on public holidays.

36 7.5x35x16/T.M. TOY/B+W/1930'S Made in the early 1930s, this car is an amalgam of advanced technology. The battery case is underneath the car, and the wheels are made of tinplate. The printed areas have been pasted onto the tram's shell, while the ceiling of the car is hand painted.

37 8.5x36x13/H. YAMADA/W/1930'S

38 29x29x9/MASUDAYA/W/1930'S   This is a model kit for a classic car. Though the package is slightly misleading, implying that the car is life-sized, it looks like it would have been a wonderful toy.

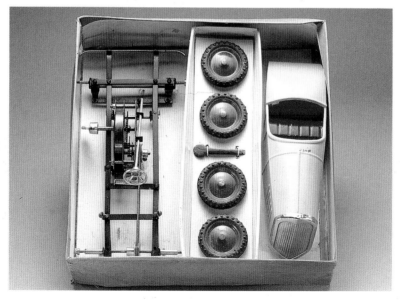

40 10x32.5x8/MASUDAYA/W/1930'S   This figure is a model of an art deco car of the thirties. When the front bumper hits a wall, it changes direction and goes backwards.

41 8x33x7/KOSUGE/W/1930'S   This is a copy of a British race car, The Bluebird. It was made in three different sizes; the one pictured here is the largest.

42 9x25x12/KONO KAKUZO/B+W/1930'S This model of a classic car was very expensive in Its day.

43 11x27x11/MASUDAYA/B+W/1930'S This model of a classic car was very expensive in its day.

44 10x28x11/UNKNOWN/B+W/1930'S The wheels of this model are tinplate, but the body of the car is made of wood. This was another expensive toy.

45 12x18x11/UNKNOWN/W/1920'S   This motorcycle features exceedingly long steering handles, a sidecar, and a very detailed engine.

46, 47 7x24x16/MASUDAYA/W/1930'S   These motorcycles are an exceptionally close copy of a German model. The sporty
clothes of the riders reveal a Western influence.

48 11x19x13/NOMURA/W/1930'S

49 10x28x16/MASUDAYA/W/1930'S This is a wind-up military three-wheeled motorcycle.

50 7x20x9/K.K. TOY/1930'S   This figure cannot move on its own accord.

51 9x27x17/ISHII/W/1930'S   This is a rather large toy, with room for more passengers in the back cart.

52 9x19x12.5/Y.C. TOY/W/1930'S   Most toy tractors were produced after World War II. This example, made in the 1930s, is very rare.

53 8x22.5x14/W.U. TOY/W/1920'S   This brightly painted
horse and cart was made in the 1920s. When it is wound up,
the back wheels move and push the horse in front of it.

54 6.5x19x12/H. YAMADA/W/1920'S   This is a copy of a
Western chariot. The pony moves by leaping up and down,
and the passenger bounces up and down in his cart.

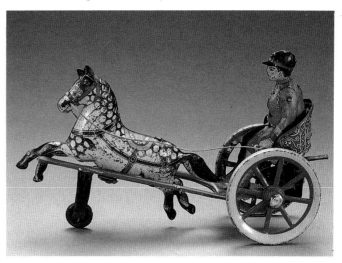

55 20.5x13x32/CHIBA/W/1920'S   This was one of the most expensive toys. When the player
strikes a ball, it flies up and out one of the holes numbered from one to six. Wire tunnels are
used to direct the balls into the holes.

56 11x8x29/UNKNOWN/W/1930'S   This peculiar figure is a Harold Lloyd doll. Varying expressions can be created by sliding the tin panels on his face into different configurations. The body also rotates from side to side.

57 7x20x12/EGAWA/W/1920'S   When this toy is wound up, the player hits the billiard balls, and they shoot into the pockets.

58 3x18x16/UNKNOWN/1920'S   This push-along bicyclist was made in the 1920s, which accounts for the natty plaid suit the cyclist wears. The toy as a whole has been made with great care.

59 11x15x18.5/SANKEI/B/1950'S
This is a baseball player who never strikes out. Once the ball is in position, he takes a swing at it.

60 10x14x61/TOPURE/B/1950'S This electrician climbs to the top of a telegraph pole with a lamp attached to his head. The electrician is, oddly enough, made up like a clown, and the entire model, telegraph pole and all, is attached to the top of the electricity company's car. This is a slightly surrealistic toy.

61 12.5x10x14/LINE MAR/W/1950'S   These models depict black musicians.

62 8x8x21.5/SUZUKI & EDWARD/W/   **63** 10x10x26.5/LINE MAR/B/1950'S
1950'S

Figures 62 and 63 are tap dancers: 62 jumps when it is wound up, and 63's
pedestal revolves via batteries.

64 10.5x17x27/UNKNOWN/B/1950'S

Figures 64 and 65 are media bears. A snakelike creature pops out of the camera lens when Figure 64 is activated. Figure 65's microphone is actually a battery case.

65 15x14x26.5/MUTSU SEISAKU/B/1950'S

66 11x17x24/SUZUKI & EDWARD/B/ 1950'S

67 11x17x24/SUZUKI & EDWARD/B/ 1950'S

68 11x17x25/SUZUKI & EDWARD/B/ 1950'S

Figures 66 through 68: The tip of the drill held by the bear dentist lights up. In Figure 67, the baby bear kicks its feet in an effort to struggle loose from the hairdresser.

69 15x16x20/SUZUKI & EDWARD/B/1950'S   This very cute toy allows a child to talk to his bear via telephone. When the child dials the phone, the bear moves his mouth as though chatting on the phone.

70 15x16.5x21/UNKNOWN/B/1950'S

71 9x19x24/UNKNOWN/B/1950'S

Figures 70 and 71: Vacuuming animals are a rather strange concept until one realizes what a novelty the vacuum cleaner was when the toy was made in the 1950s

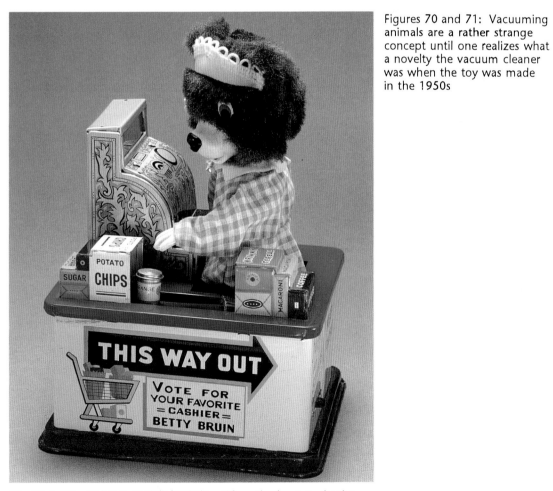

72 13.5x15.x20/LINE MAR/B/1950'S When the bear works the cash register, numbers appear. Toy cartons and fruit flow by as though they were on a conveyor belt.

73 18.5x18.5x17.5/ WAKASUKO BOEKI/B/1950'S When the father bear catches a fish, he drops it into the bucket held by the baby bear.

74 10x16.5x21.5/ALPS/B/1950'S A device is hidden under the water at the bottom of the cup to create a wavelike effect on the surface, which makes the goldfish move. The result is to make the cat look as though he were playing with something. The cat's eyes light up and its tail lashes about in a very catlike fashion.

75 8x13x11/UNKNOWN/W/1950'S   The awkward placement of this bear's legs and arms conjures up the image of what a real bear would look like riding a motorcycle.

76 14.5/15.5/22./UNKNOWN/B/1950'S   Both the canvas and palette of this artist monkey light up.

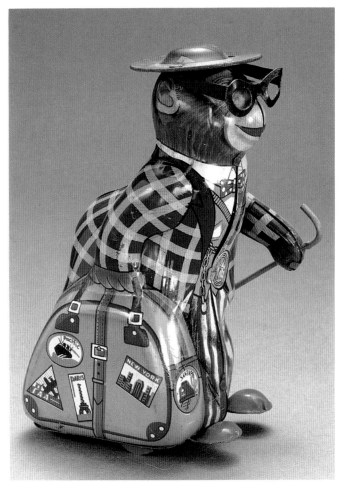

77 7x7x12/SANKEI/W/1950'S   The shape of this toy is unususal, but its clothes, walking stick, and watch are all depicted in great detail.

78 12x24x12/EXELO/B/1950'S

79 7x20x13/UNKNOWN/F/1950'S

80 7x21x11/UNKNOWN/F/1950'S

81 14x27x17/YONEZAWA/F/1950'S

Figures 82 through 104: One of the most popular subjects for tin toys is circus characters, especially clowns.

82 6x24x12.5/MITSHUHASHI/F/1950'S

83 6x26x9.5/YOSHII/F/1950'S

84 7x23x10/SANKEI/F/1950'S

85 6.5x30x9/UNKNOWN/F/1950'S

86   8x24x8.5/KANTO/F/1950'S

87   6x26x9/NOMURA/F/1950'S

88   8.5x52x13/SANKEI/F/1950'S

89   6.5x23x14/MITSUHASHI/F/1950'S

90   11x6x17/NOMURA/W/1950'S

91 9x7x22/UNKNOWN/W/1950'S

92 10x15x27/SUZUKI/B/1950'S

93 7x13x23/KURAMOCHI/W/1950'S

94 8.5x9x21/ALPS/W/1950'S

95-98 7x10x27/UNKNOWN/W/1950'S

99–101 8.5x11x21/NONURA/W/1950'S

102–104 8x9x26, 8x5x19/UNKNOWN/W/1950'S

Figures 105 through 107 are variations on models of B-50 jets. On some, the propeller lights up, and on others, it actually whirls. All of these are large toys.

105   48x37x14/TOMIYAMA/B/1950'S

106   48x37x14/TOMIYAMA/F/1950'S

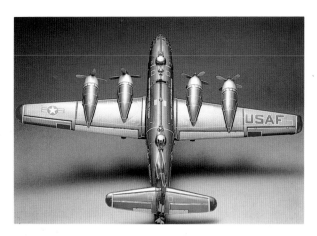

107 48x37x14/TOMIYAMA/B/1950'S

108 40x32x8/KOKYU SHOKAI/F/1950'S

109 60x46.5x16.5/YONEZAWA/F/1960'S

110 44.5x34x13.5/BANDAI/F/1960'S

111 36x28.5x11/HADSON/B+F/1950'S

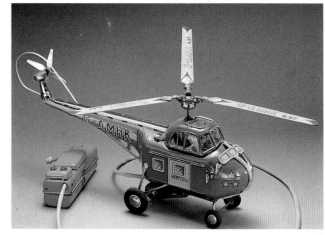

112 28x35x16/UNKNOWN/B/1950'S

113   45x43x17/TOMIYAMA/F/1960'S

114   24x28x13/TOMIYAMA/F/1960'S

115, 116   19x24x10, 16x26x10/TC MIYAMA/NOMURA/F/1960'S

117 60x55x16.5/YOZEZAWA/B/1960'S

118   36x28.5x11/YONEZAWA/F/1960'S

119   60x55x16.5/YONEZAWA/B/1960'S

120   49x64.5x19.5/BANDAI/F/1960'S

121   48x37x15.5/NOMURA/B/1950'S

122   48.5x40x12/ASAHI/F/1950'S

123, 124 32x23x8/UNKNOWN/F/1950'S, 1940'S

125 13.5x11x8.5/NOMURA/W/1950'S

126, 127 25x18x10/BANDAI/F/1950'S

128 25x22x10/KOKYU SHOKAI/W/1940'S

Figures 129 through 131 are NBC broadcasting trailers. Figure 129 is battery operated; when it is switched on, a broadcast starts playing on the TV screen on the side of the bus.

129   9x21x17/YONEZAWA/B/1950'S

130   7.5x20x11/UNKNOWN/F/1950'S

131   5x16x10/UNKNOWN/F/1950'S

132 9x30x17/LINE MAR/B/1950'S

134 12x31x17/ICHIKO/B/1950'S

133 9x24x26/MARUSAN/B/1950'S

135 10.5x24x11/UNKNOWN/F/1960'S

136 8x24x8.5/BANDAI/F/1950'S

137 8x32x10.5/KANAME SANGYO/F/1950'S

138 8x32x10.5/KANAME SANGYO/F/1950'S

139 8x32x10.5/KANAME SANGYO/F/1950'S

140 8x33x11/LINE MAR/F/1950'S

141 26x15x8.5/UNKNOWN/F/1950'S

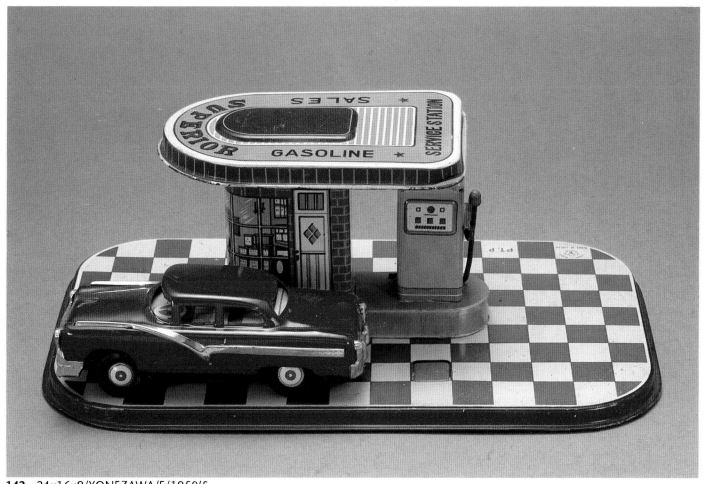

142 24x16x9/YONEZAWA/F/1950'S

143 9x32.5x12/MARUSAN/F/1950'S

144 8.5x26x10/NOMURA/F/1950'S

145 5x16x8/MASUDAYA/F/1950'S

146 6.5x19x8/TOMIYAMA/F/1950'S

147 13x22x13/YONEZAWA/B/1950'S

148 12x25x15/NOMURA/B/1950'S

149, 150 9x18x13, 9.5x28x10/YOSHIYA, KANAME SANGYO/F, B/1950'S

152 12x28x20/NOMURA/B/1950'S

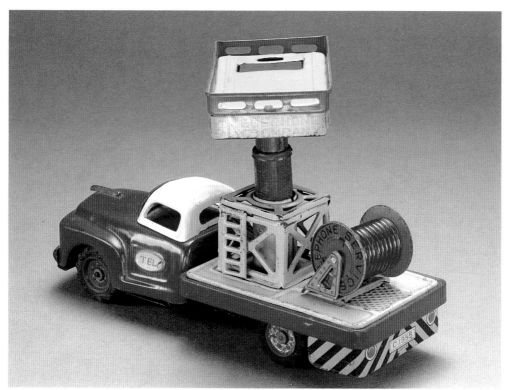

151 7x19x13/UNKNOWN/F/1950'S

153 8.5x19x8/MITSUHASHI/F/1950'S

154 15.5x49x14/MARUSAN/F/1950'S

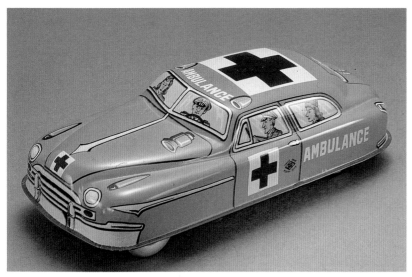

155 10.5x23.5x7/MASUDAYA/F/1950'S

156 8x21x7/BANDAI/F/1950'S

Figures 157 through 159: Fire engines, once very popular with children, were produced in great number in the 1950s.

157 6x22x10/NOMURA/F/1950'S

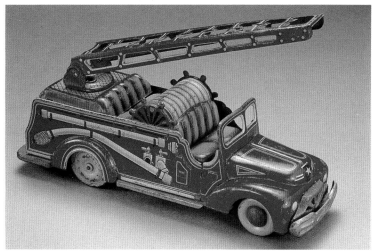

158   7x23x9/UNKNOWN/F/1950'S

159   9.5x26x11/UNKNOWN/B/1950'S

160   10x20x8/ALPS/F/1950'S

Figures 161 through 165: These are imaginative renditions of the San Francisco cable car. Figure 162 is a wind-up. Its seats and wheels have been painted in great detail. The refined color scheme would have made it possible to use this toy as a decoration.

161 7x21x12/UNKNOWN/F/1950'S

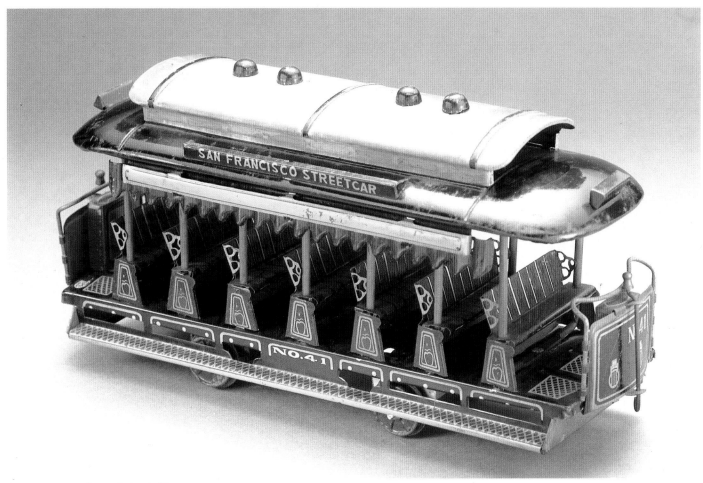

162 8x25x11/ALPS/F/1950'S

163 5.5x18x8/HISHIMO/F/1950'S

164 6x15x8.5/NOMURA/F/1950'S

165 6x15x8.5/HISHIMO/F/1950'S

166, 167 12x53x19, 10x52x16/YONEZAWA, MARUSAN/F/1950'S

168 12.5x53x16/YONEZAWA/F/1950'S

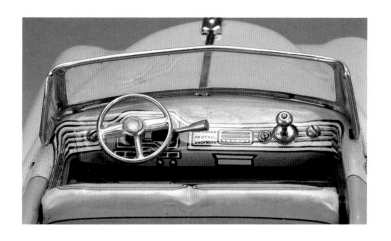

169 15x36x12/NOMURA/F/1950'S   This Cadillac's batteries are located under the seats. It moves forward and backward by means of a lever. The car's headlights also work.

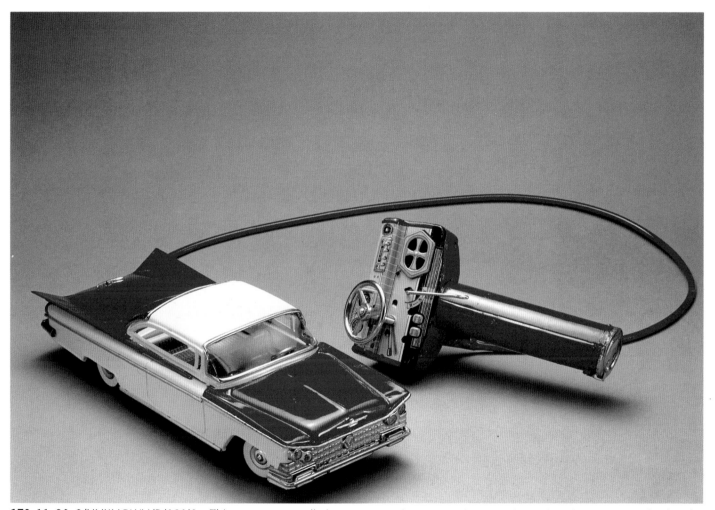

170 11x30x9/UNKNOWN/B/190'S   This remote-controlled car operates by means of a control panel with a steering wheel and various levers. It can also move backwards.

171, 172 21x41x13/NOMURA/F/1950'S These are the futuristic toy cars of the 1950s. At the time, they must have seemed to be masterpieces of technology.

173, 174 12.5x32.5x10/YONEZAWA/F/1950'S   These friction-operated Lincolns were made in the 1950s.

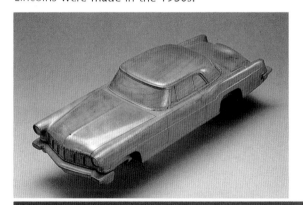

175, 176 10.5x29x8/YONEZAWA/B+F/1950'S   These are typical examples of toy cars made in the 1950s. Thousands of them were produced, and they were always modeled on American cars.

177 17x40x12/ALPS/F/1950'S

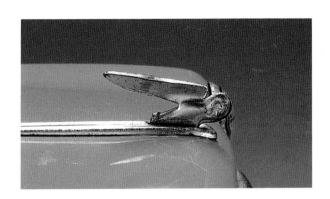

178 17x40x12/ALPS/F/1950'S

179 10.5x25x8/YONEZAWA/F/1950'S

180 15x39x10/ASAHI/F/1960'S

181 15x39x10/ASAHI/F/1960'S

182 20x66x18.5/YONEZAWA/F/1950'S   This is the largest of all toy cars made in the 1950s.

183 5x20x15.5/MARUSAN/F/1950'S   This toy was made during the renaissance of the scooter.

184, 185 8x24x16/YONEZAWA/F/1950'S

187 5x18x13/UNKNOWN/F/1950'S

186 10x28x19/YONEZAWA/F/1950'S

188 8.5x23.5x12/BANDAI/F/1950'S

189, 190 5.5x18x12/UNKNOWN/LINE MAR/F/1950'S

191 8x26x17/HADSON/F/1950'S

192 8x26x17/HADSON/F/1950'S

193 14x19x14/UNKNOWN/F/1950'S

194–196 10x30x17.5, 6x18x12/I.Y. METAL TOYS/F/1950'S

197 9x13x9/NIKKO KOGYO/W/1950'S

198 8x30x20/MASUDAYA/B/1950'S

199 5x15x7.5/UNKNOWN/W/1950'S

200, 201 7.5x10x6/UNKNOWN/F/1950'S

202 6x13x8.5/KOSHIBE/F/1950'S

203, 204 5x15x10/UNKNOWN/F/1950'S

205 5x16x9/UNKNOWN/F/1950'S

206 5x12x9/UNKNOWN/F/1950'S

207 9x30x19/I.Y. METAL TOYS/F/1950'S

209 5x15x7/SANKEI/F/1950'S

210 8x29.5x10/LINE MAR/F/1950'S

211 10x33x10/UNKNOWN/B/1950'S

212 9.5x32x10/NOMURA/F/1950'S

213 8x35x11/MASUDAYA/F/1950'S

214 8.5x25x9/BANDAI/B/1950'S

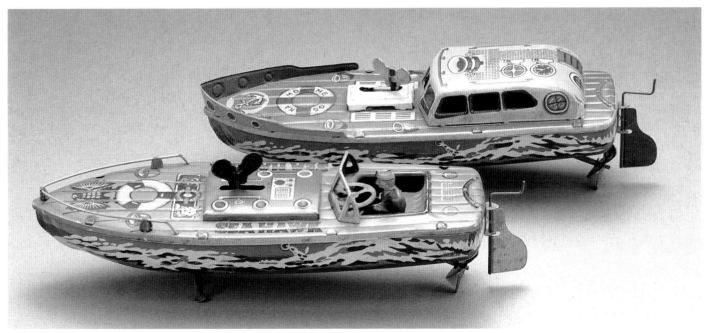

215, 216 7.5x25x8/UNKNOWN/W/1950'S

217 29.5x11x29/ASAHI/B/1950'S

ENCOUNTERING TIN TOYS

My collection of tin toys grew out of the toys I had as a child, and it reveals the great impact that American style and customs had on Japanese children in the fifties. To me, the American cars that ran through the streets exuding gaudy wastefulness were incomparably smart and fascinating. American dramas and cartoons kept me glued to the television. I dreamed of dressing in the tweed jacket with leather elbow patches worn by Robert Young on his television program. This obsession with things American must have developed because Japan in the 1950s was politically influenced by Europe and America, particularly the latter. In childhood, we yearned for "America" without understanding all this social logic. The result was that the tin toys made in Japan at that time were crammed with the yearning and astonishment the people of Japan then felt for American culture. With their simple design, coloring, and movement, tin toys continue to fascinate me, and even now, they convey an unchanged innocence and serenity.

—Teruhisa Kitahara

YESTERDAY'S
TOYS
Robots, Spaceships, and Monsters

YESTERDAY'S TOYS

Robots, Spaceships, and Monsters

Teruhisa Kitahara ▪ Photography by Masashi Kudo

YESTERDAY'S TOYS—OBJECTS OF WONDER

Robots and other science-fiction-related toys were always objects of wonder to children. Even the simplest spring-operated robot looks powerfully mechanical to young eyes.

The word *Robot,* referring to an artificial humanoid, first appeared in the drama *R.U.R.,* written by Czechoslovakian author Karel Capek in 1920. But it is not clear when robots made their first appearance as toys. There seem to have been mobile, doll-like toys for quite some time, but these were not actual robots.

Toy robots reached their height of popularity from the 1940s to the 1960s. This can be said for tinplate toys in general; they all experienced their peak during this period. The archetypal robot of the era was Robby. Robby was introduced in 1956, when he appeared in Metro-Goldwyn-Mayer's science-fiction movie *The Forbidden Planet.* This film tells the story of a scientist who was shipwrecked for years on another planet, where he created a robot called Robby, a machine with a human personality, who could speak human language, had feelings, and could even eat if he wanted to. From this character, the toy Robby was created. Although toy robots varied in size, means of movement, and materials, each robot had some of the characteristics of Robby.

Another famous robot was Goto, who was introduced in the Twentieth Century Fox movie *The Day the Earth Stood Still.* After these two prototypes, many other toy robots and spaceships became popular. There were a variety of kinds, from simple toys joining square parts of different sizes, to complicated toys with flashing lights and missile launchers on their fronts. The era of robot and science-fiction toys was also the era when manufacturing techniques were progressing dramatically. Not only were there advances in mechanical techniques, but even the methods of printing on tin changed the world of toy robots, allowing the tiniest gauges and meters to be printed in detail.

In many of these toys, especially those made in Japan, we can see the influence of television and comic heroes such as Steelman and Astroboy. Many variations of these heroes were produced. Sometimes, to cut down production costs, robots were cast in the same molds and changed only by small printing details into different toys.

Robots and science fiction toys have undergone many changes and have now been surpassed in popularity by television and video games. But their futuristic shapes and materials have the perennial fascination of the new. The wonder and excitement of the unknown are always with us.

1–4 15x33x22.5, 17.5x13.5x31, 9x6.5x16/NOMURA, YOSHIYA/B,F/1950'S

5 15x33x22.5/NOMURA/B/1950'S (battery operated) This is "Robby," which first appeared in the 1956 film The Forbidden Planet. In Figure 5, Robby appears in combination with a space patrol car. He can be seen on his own in Figure 6.

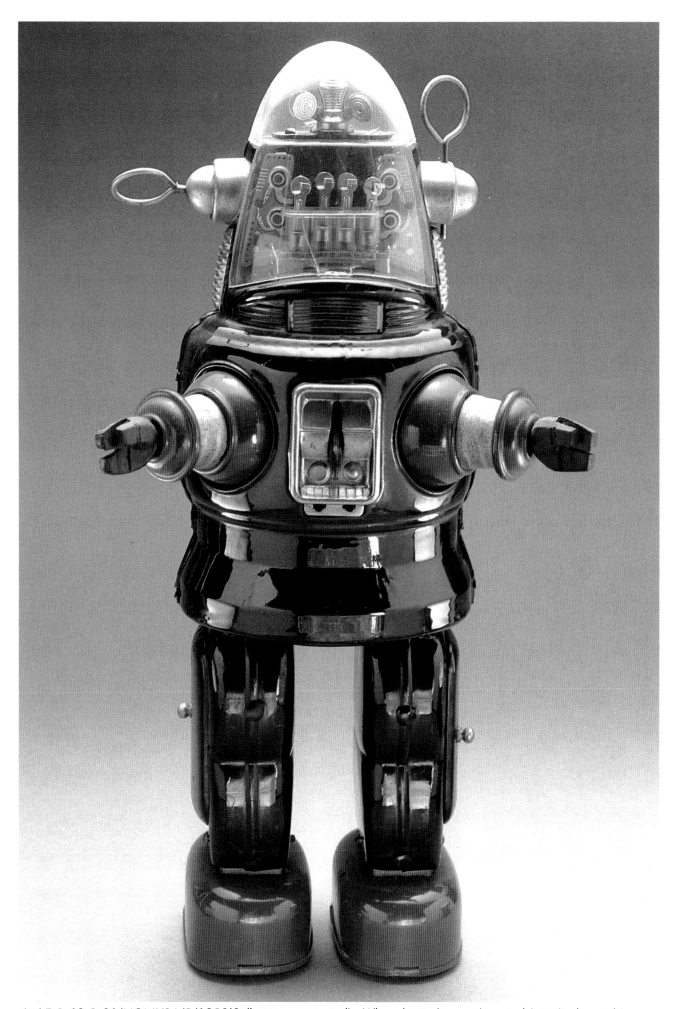

6 17.5x13.5x31/NOMURA/B/1950'S (battery operated) When batteries are inserted into its legs, this robot walks and the piston on its head goes up and down. After this Robby robot appeared, many robots were made with the same trademark transparent head.

7,8 9x6.5x16/YOSHIYA/F/1950'S

9 9x6.5x16/YOSHIYA/B/1950'S

10–12 7x5.5x16/YONEZAWA, LINE MAR/W, B/1950'S

Figure 11 (battery operated): The sci-fi transparent helmet is the most distinctive feature of this toy. This 1950s, battery-operated robot is extremely rare. Figures 10, 11, and 12 are variations on the same toy.

13, 14 13.5x10x33/ROSKO/B/1950'S (battery operated) The hands and feet of these robots move when batteries are inserted into their feet. Each robot holds a transceiver and wears a panel of red lights that turn on and off. These light switches also operate lights that illuminate the faces of the robots.

15, 16 7x7x19/NAITO SHOTEN/W/1950'S (wind-up) This robot was produced in the 1950s. The coiled wire pipe leading from its head to the tank on its back gives it a futuristic, sci-fi feeling. The expression on this robot's face is very interesting.

17 9x14.5x21/NOMURA/1950'S (battery operated) It is very unusual for a robot to play a musical instrument; at best, a robot can only manage a drum. This robot's hands move up and down to strike the drum and its feet march and its eyes light up.

18 7x8x19/NOMURA/W/1950'S (wind-up) Dressed as an astronaut, this robot carries a space can on his hips. Robots without helmets on their heads are quite common, but this one, with his head encapsulated in a transparent dome, is unique. Both hands and feet are movable.

19 11x5.5x16.5/LINE MAR/B/1950'S (battery operated) This robot's eyes light up, and pins can be extended from its feet to enable it to walk. This kind of robot with walking pins was very common in the 1950s, but is now very rare.

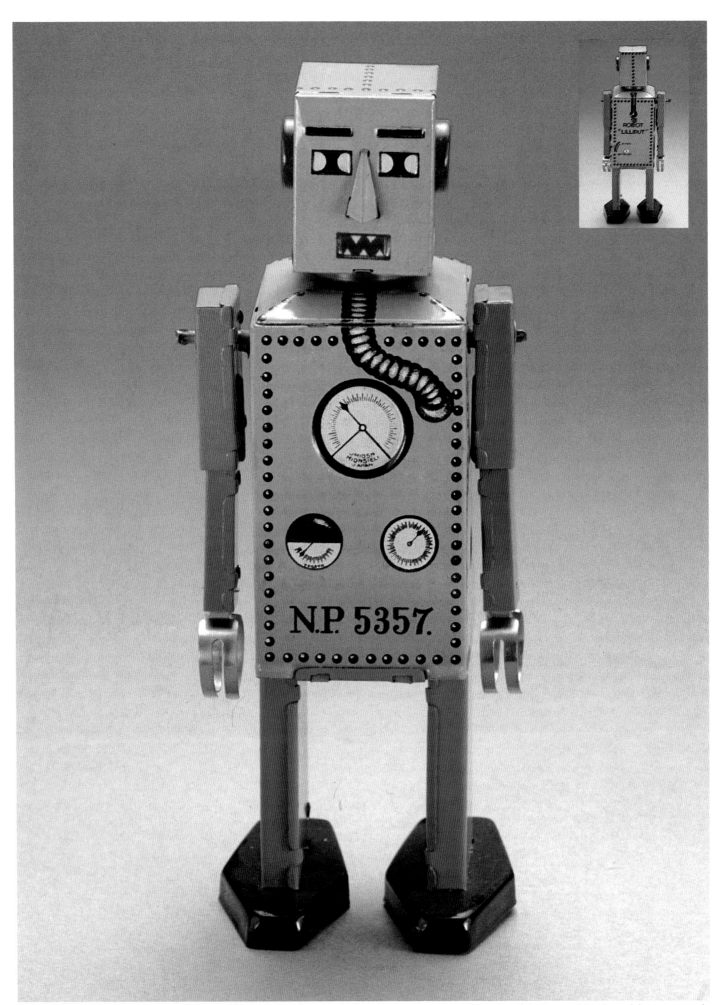

20 6.5x5x15.5/UNKNOWN/W/1940'S (wind-up) This robot was constructed in the 1940s and is the oldest example of its kind. Its boxy head and body give it a primitive, simplistic look.

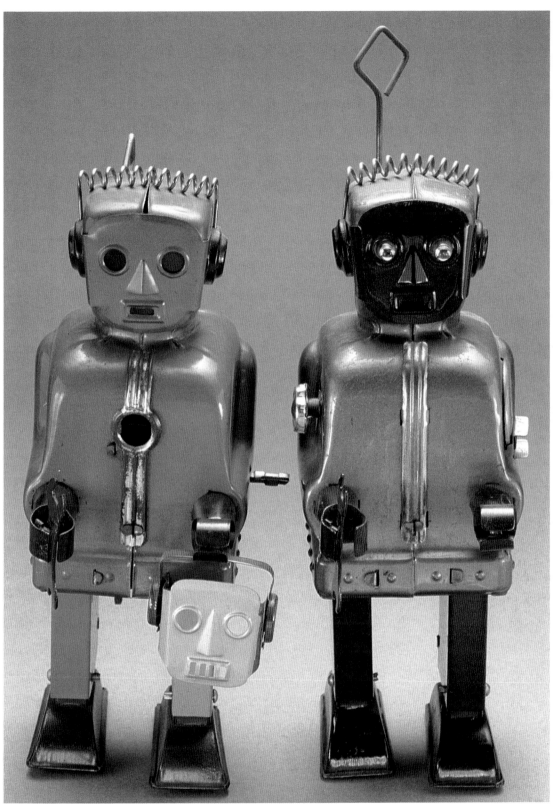

21, 22 10x7x19/NOMURA/B,
W/1950'S (Figure 21, wind-up,
and Figure 22, battery operated)
Apart from their difference in
mechanism, these robots are quite
similar. The clamp-style hands
indicate that these are powerful
robots, able to grip as well as
walk.

23 7x10x24/NOMURA/B/1950'S (battery operated) Embellished with a parabola antenna on its head and eyes that light up, this amusing robot was apparently a bestseller, as it went through a variety of new packages in its considerable lifetime.

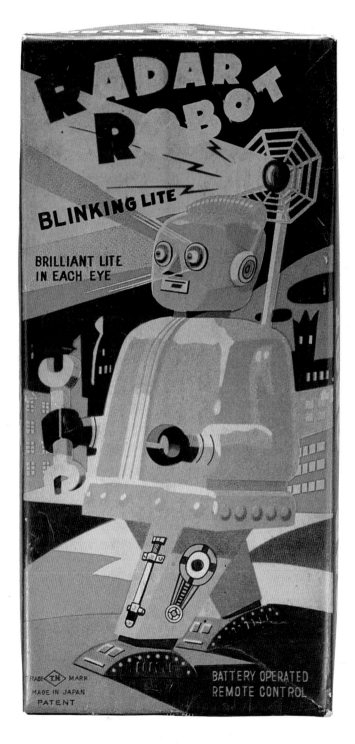

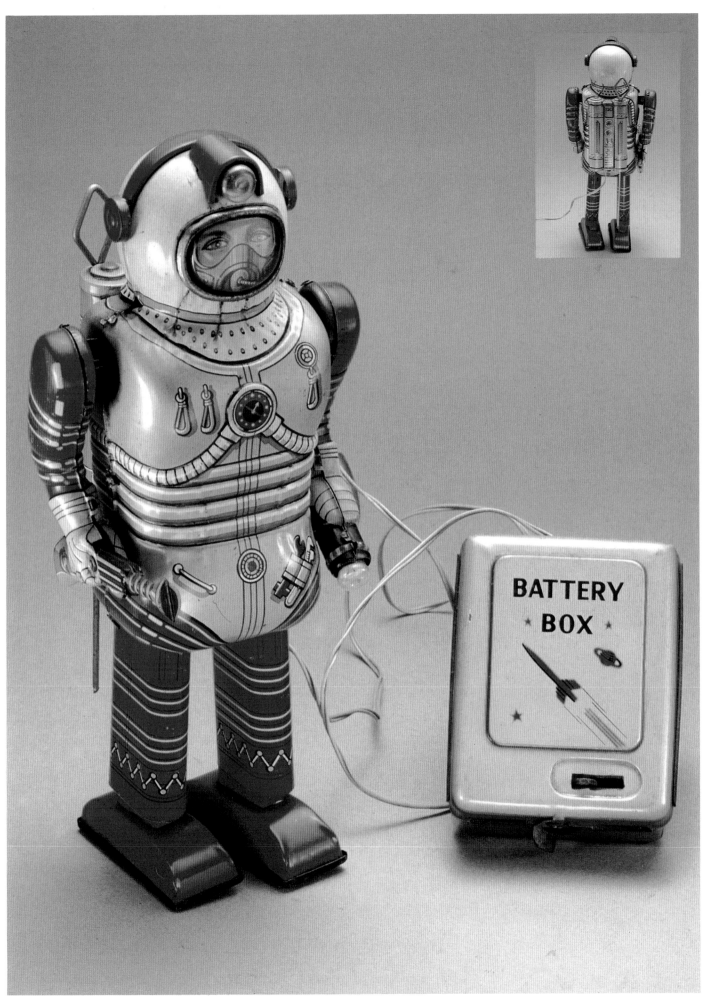

24 10x8x22.5/NOMURA/B/1950'S (battery operated) Combining the costume of an astronaut with the perambulation and lighting effects of a robot, this futuristic toy was produced around the time that Yuri Gagarin was the first man in space.

25 11.5x6.5x19/MASUDAYA/B/1950'S (battery operated) This robot has a variety of functions—it moves both forward and backward, its head lights up, and its hands move. Batteries for Figure 26 (right) are inserted into the tank on its back, and this weight is countered by lead on its front. Its antenna serves as its on-and-off switch.

26 11.5x6.5x19/MASUDAYA/B/1950'S

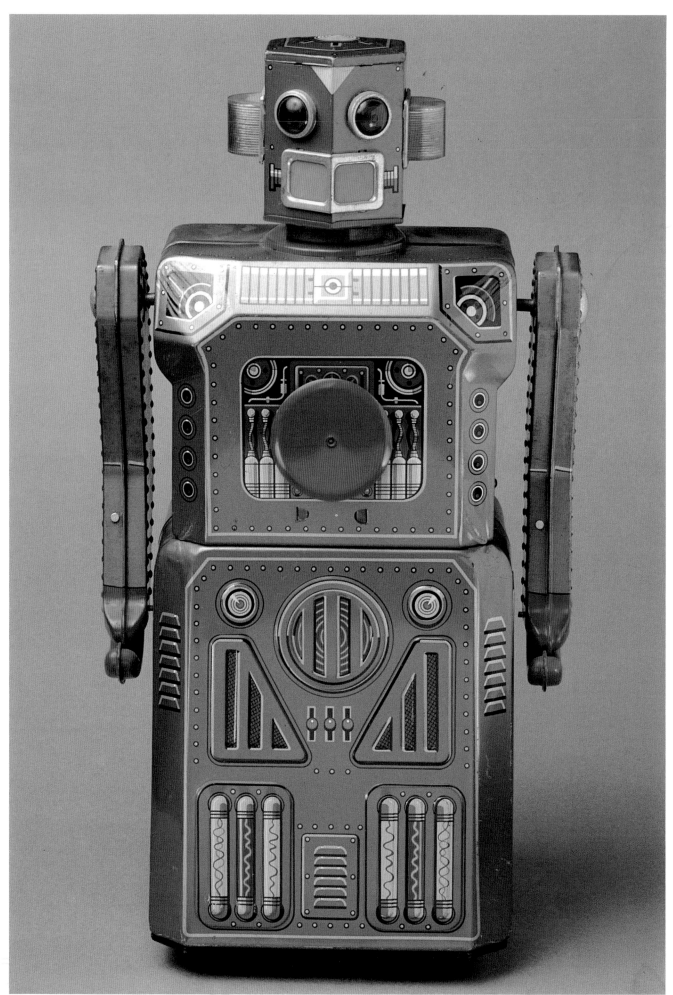

27 21.5x15.5x38/MASUDAYA/B/1950'S (battery operated) This robot has no feet. When the circle on its chest touches a rubber suction disk beneath it, the wheels below go into action. A light inside its mouth illuminates its head, and its bulky monumentality gives an impression of power.

28 21.5x16x38/MASUDAYA/B/1950'S (battery operated) Another footless robot, this example has lights in its eyes, head, and mouth and a siren that goes off. Like Figure 27, its monolithic shape makes it seem quite powerful.

29 16.5x11x45/BANDAI/B/1960'S (battery operated) Inserting an antenna operates this robot's switch. The robot fires missiles from its chest every five or six steps. The missiles are surprisingly large.

30, 31 16.5x11x45/BANDAI/B/1960'S

32 17x10x32/UNKNOWN/B/1950'S (battery operated) After taking a few steps, the face of this robot splits open to reveal the face of an angel. The head then joins itself together and it continues walking, only to repeat the performance again a few steps later with the face of a devil. This is a very peculiar robot.

33 9x8x22/YONEZAWA/W/ 1950'S (wind-up) This odd-looking robot has wrenches for hands and can walk forward while his arms remain stationary. Figure 36 has the same body with a different head.

34 8.5x8.5x19/UNKNOWN/1950'S (wind-up) This example is the most robot-like of all. It is so typical that it is difficult to enumerate its distinctive features.

35 7.5x7x18/UNKNOWN/F/1950'S (friction operated) This friction robot is operated with the use of a crank. A push will send it moving forward, waving its head from side to side.

36 7x8.5x21.5/YONEZAWA/W/1960'S (wind-up) This robot is simply a variant of Figure 33, but it's obvious from the package that it was based on a children's television program called "The Jet Boy." Sparks are emitted from its eyes by means of a flint.

37, 38 17x7x24.5/YONEZAWA/W/1950'S (wind-up) These strange robots were designed with a telephone theme—their arms are telephone receivers and dials appear on their chests. The receivers have been made from two different materials: tinplate and plastic.

39 12x10.5x24/YONEZAWA/W/ 1950'S (wind-up) Similar in shape to Figures 37 and 38, this robot has abandoned the telephone motif for a variety of other features, including an oxygen meter and hologramic eyes that seem to wink when seen from an angle.

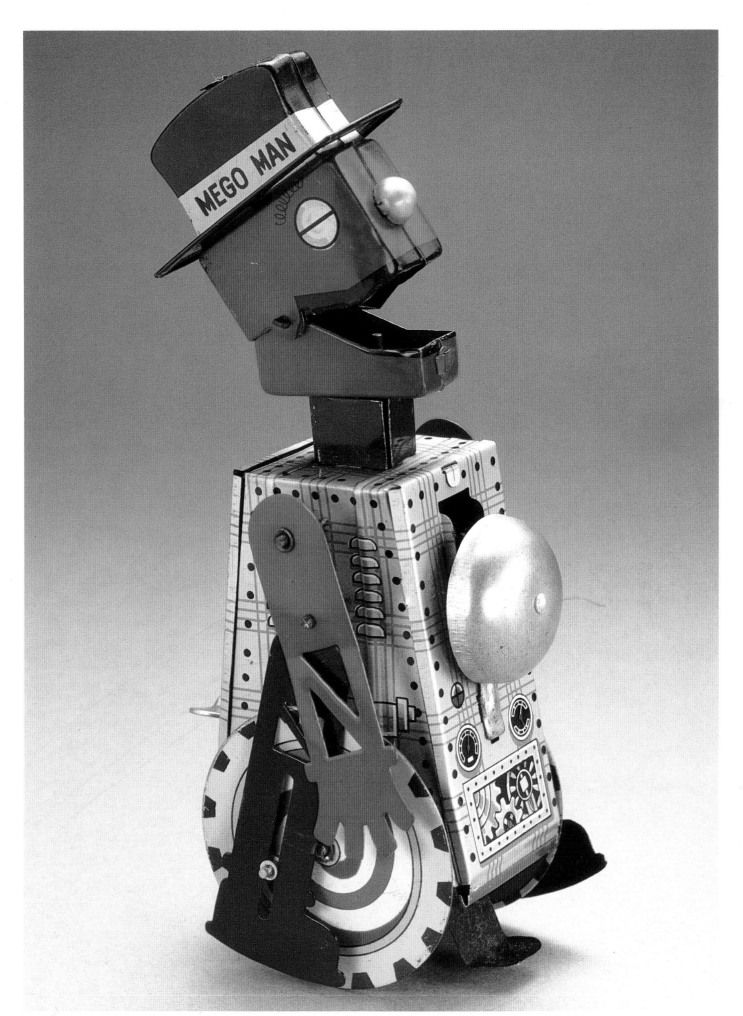

40 6.5x7x17.5/S.Y. TOY/W/1960's (wind-up) This example is half-human and half-robot. The design has been strongly influenced by the television program "Robot Santohei (the third private)," which was popular when it was made. It opens and closes its mouth continually while its bell rings.

41 8.5x6x14.5/TOMIYAMA/
W/1950'S (wind-up) This robot
has a cute face and distinctive
feet that resemble an ice-shaving
machine. The new-look plastic
legs contrast with the old-style
body.

42 15x11.5x27.5/YONEZAWA/
B/1950'S (battery operated) When
it bumps into a wall or other
object, this robot turns around and
advances away from the
obstruction. Its head can be
removed to replace the lightbulb
that illuminates its eyes.

43 16x10x31/UNKNOWN/B/1950'S
(battery operated) This novelty robot
blows smoke out of its mouth and has a
siren on its head.

44–46 16x10.5x30/YONEZAWA/B/1950'S (battery operated) Every few steps, these robots will stop, flash their eyes on and off, and then blow smoke out of their mouths. The smoke is produced by a wire inside being heated and reacting against some oil-soaked cotton wool. The smoke is then blown out of the head by a bellows.

47–49 8x6x14, 9x8x21, 8.5x6x15/KOGURE/B/1960'S (battery operated) These robots flash their eyes while walking. All the parts are exceedingly well expressed considering that they are made of plastic.

50 10x10x40/MASUYA/F/1960'S (friction operated) When the point of this horizontal rocket comes into contact with any form of obstruction, a lever is operated, and it jerks up to a vertical position. Two flights of steps can be attached to allow easy entry for the astronauts.

51 18.5x39x14.5/YONEZAWA/B/1950'S (batt
look like. The cockpit lights up, and various
opening the compartment near the front whee

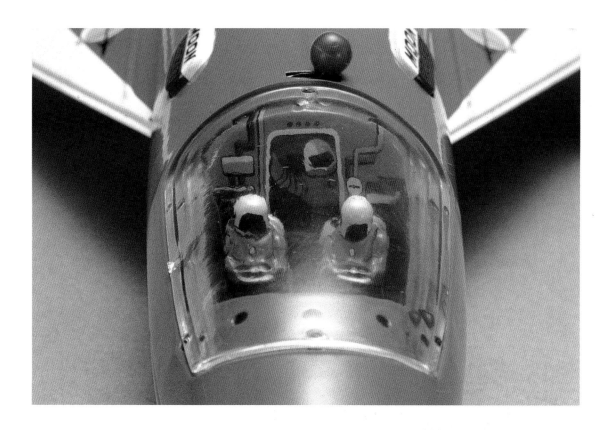

operated) This is the 1950s idea of what the space shuttle of the future would
devices, such as an antenna, can be attached. The lightbulb can easily be changed by

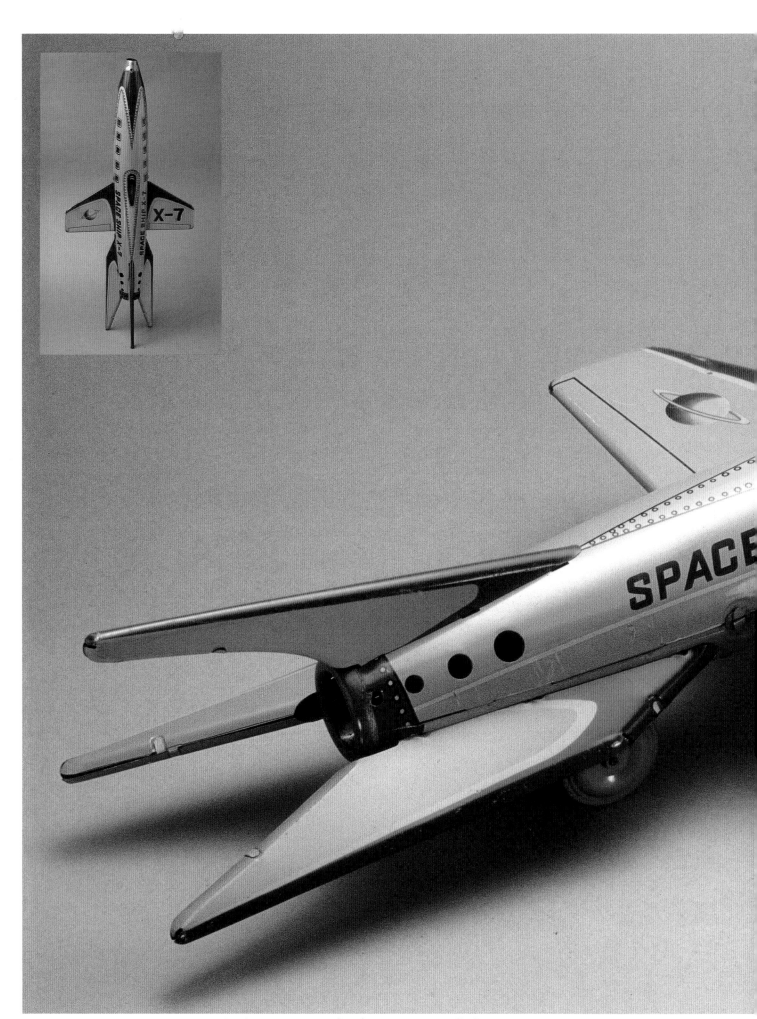

52 26x56x11/MASUDAYA/F/1950'S (friction operated) This is an unusually large rocket for its type. Such a huge toy mus[t]
have been a delight for small children.

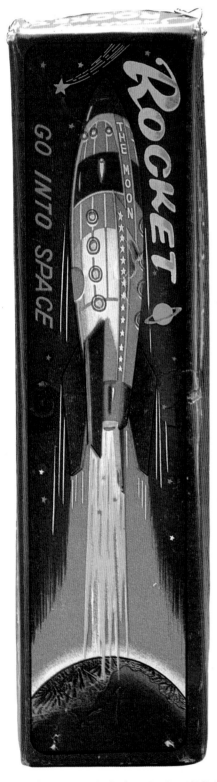

54 7X7X32/
UNKNOWN/F/1950'S

Figure 54 and 55 (friction operated): This 1950s toy only moves forward on its wheels. Its mechanical simplicity has been counterbalanced by its colorful design and interesting shape.

53: 5.5x5.5x24/MARUSAN/F/1950'S This uncommon rocket is hung by a string. The nose of the rocket is hooked onto something high, and then the body is pulled downward. When it is released, the rocket will rewind itself on the string and make sounds like a blast-off.

55 7X6X31/MASUDAYA/F/1950'S

56 10x45x12/UNKNOWN/B/1960'S (battery operated): The front and back ends of this rocket light up. The lid of the cockpit opens to reveal the pilot.

Figures 57 and 58 (wind-up): These are American toys from the 1940s. Figure 57 seems to have been taken from some sort of story, while Figure 58 features a character from the omic strip "Buck Rogers." Sparks fly out the back as the rocket moves noisily forward.

57 8x31x9/MARZ/W/1940'S

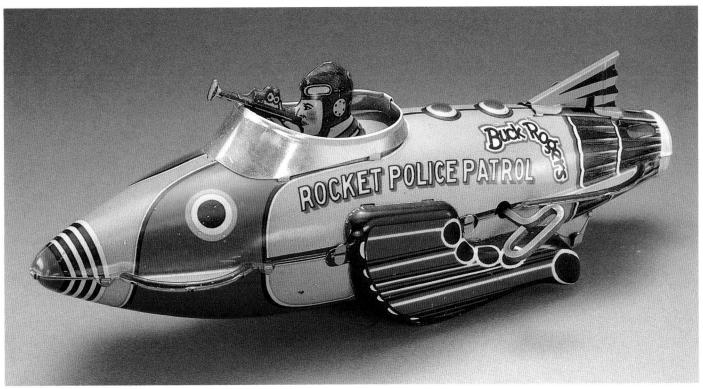

58 9x31x12/LOUIS MARX/W/1940'S

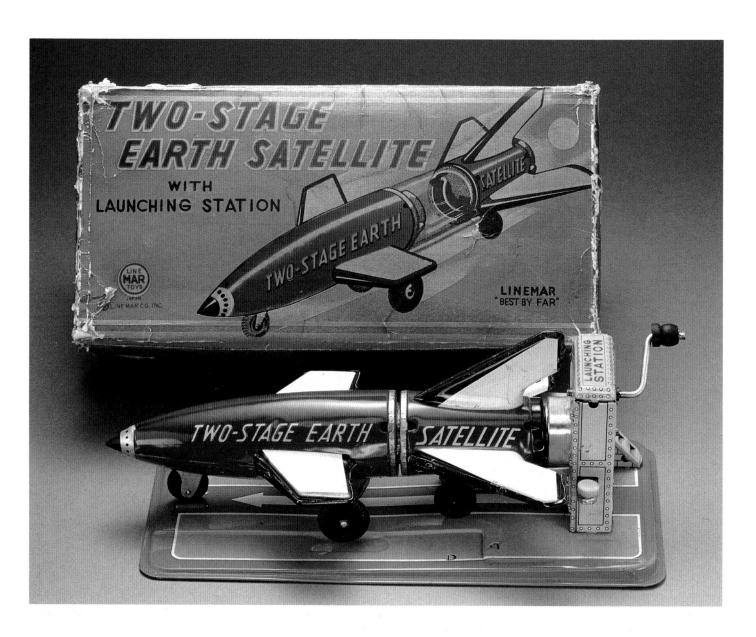

59 12x26x1/LINE MAR/F/1950'S (friction operated): Friction for this rocket is generated by winding the crank and then releasing the rocket. One of its interesting features is that halfway through its journey, it jettisons its booster rockets. Although it was designed for ground use, it creates a perfect outer-space atmosphere.

60 (battery operated): This simple rocket was created in the 1950s; although it only moves forward and lights up, its body pattern is beautiful. Its package gives it an extra sci-fi touch.

Figures 61 and 62 (battery operated): The globe of Figure 61 rotates while the satellite rises on a cushion of air. The effect is that of a satellite orbiting the earth. Figure 62 is a bank; the rocket orbits the earth every time a coin is inserted.

61 13x11x23.5/S.N.TOY/B/1950'S

62 13.5x 23/WAKASUTO BOEKI/B/1950'S

63, 64 16x16x26/YONEZAWA/B/1960'S

Figure 63 and 64 (battery operated): These toys were made in the 1960s. The satellites orbit the moons.

65 13x13x38/WACO/B/1950'S (battery operated): This is a cylindrical space station. It moves forward, the antenna rotates, and a platform emerges. The light at the bottom flashes beautifully.

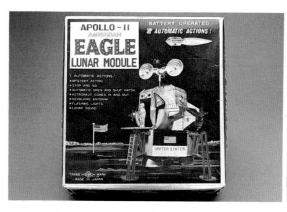

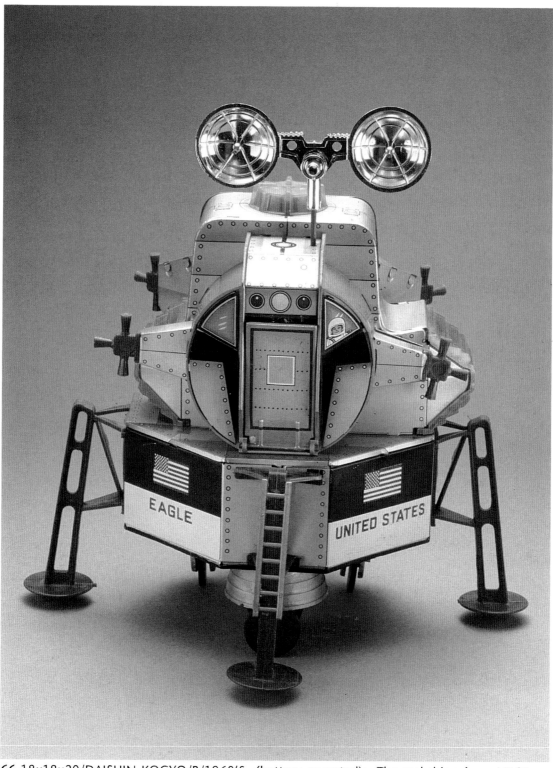

66 18x18x20/DAISHIN KOGYO/B/1960'S (battery operated): The real thing from NASA was the model for this 1960s toy. The door opens and a stairway for the astronaut appears.

Figures 67 and 68 (friction operated): The Gemini space program provided the model for these toys. The space walk is simulated by the astronaut hovering over the Gemini. Each astronaut displays a different facial expression.

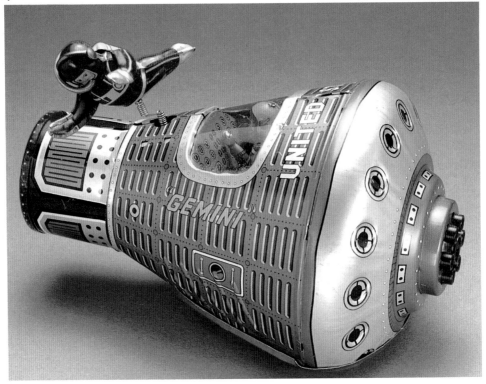

67 14x20x15/HORIKAWA/F/1960'S

68 9.5x16x11/KANTO TOY/F/1960'S

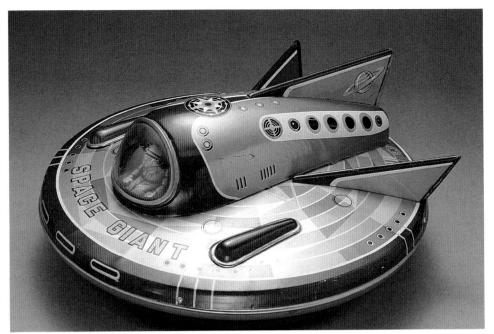

69 31x31x12/MASUDAYA/B/1950'S

70 30x30x23/HORIKAWA/B/1950'S

71 29x29x16.5/HORIKAWA/B/1960'S

72 24x24x11/NOMURA/B/1960'S

73 14.5x14.5x8.5/YOSHIYA/F/1960'S

74 14x14x9.5/UNKNOWN/F/1950'S

75 13x13x8/ASAHI/F/1950'S

76,77 12x24x22/YONEZAWA/B/1950'S

Figures 76 through 78 (battery operated): The astronauts in these examples shake their heads from side to side as they drive. The ball that rises on a jet of air behind Figure 78 demonstrates the weightlessness of outer space.

78 10x21.5x11.5/MASUDAYA/B/1950'S

Figures 79 and 80 (battery operated): This space module pulls a car containing batteries behind it. The television camera mounted on the front of the vehicle is also a monitor with changing pictures.

79 12x27x13/NOMURA/B/1950'S

80 12x27x13/NOMURA/B/1950'S

81 13x23x12/NOMURA/b/1960's (battery operated): This "space patrol car" is based on the Volkswagen. Elaborate details, including a coil for a bumper, were added to give an outer space effect. This toy was probably aimed at Volkswagen fanatics.

Figures 82 and 83 (friction operated): The rubber dolls on these space motorcycles are rather worn, but Superman is still identifiable due to his cape.

82,83   7x30.5x14/BANDA/F/1950'S

84   7.5x17x6/UNKNOWN/F/1960'S

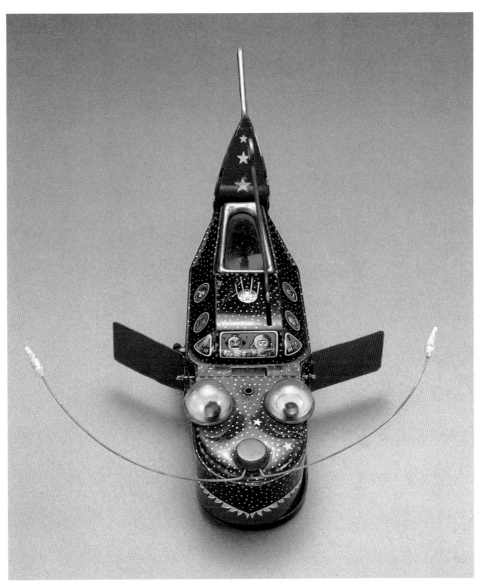

85 7x22x11/YOSHIYA/F/1950'S

86 6.5x29.5x7.5/MARUSAN/F/1950'S (friction operated): This is a very unusual combination of robot and submarine. The two entities do not seem to go together.

Figures 87 through 89 (friction operated): A sci-fi feel was achieved here by putting astonauts in convertibles or jeeps.

87 10x17.5x9.5/UNKNOWN/F/1950'S

88 8x21x8/UNKNOWN/F/1950'S

89 11x25x12/UNKNOWN/F/1950'S

90-107 SPACEGUNS/1950'S-1960'S

Figures 90 through 110: These are water pistols and battery-operated spark pistols. The battery-operated pistols produce sparks and noise, but none of them fire bullets. The water pistols have a sci-fi style.

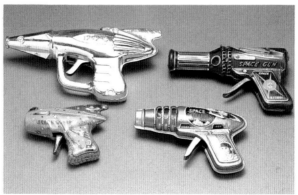

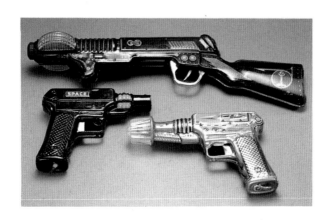

108 2.5x10x8/NOMURA/1950'S

HITS IT'S MARK!
LEAVES A MARK

X

X 100 MYSTERY DART GUN

LANYARD HOLE

GUN LO...

1—Pull
2—Insert dart and release
3—Gun is now loaded to leave its mark

EVERYTHING
IS A TARGET !

REFILL DARTS
WITH HARMLESS
TALC POWDER

LOADED FOR
500-ACTION SHOTS

No. X 100

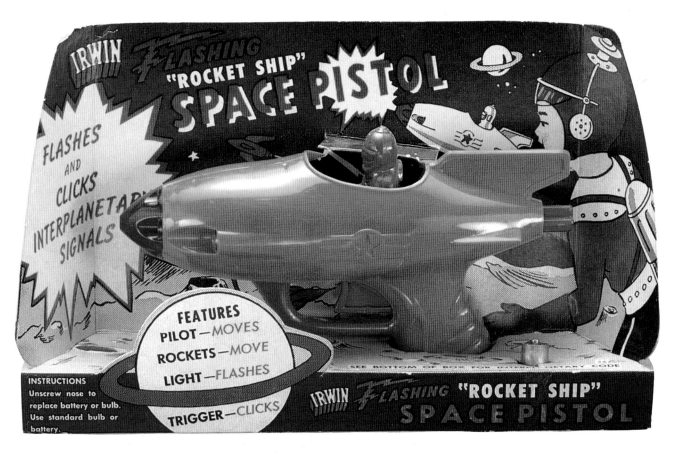

IRWIN FLASHING "ROCKET SHIP" SPACE PISTOL

FLASHES
AND
CLICKS
INTERPLANETARY
SIGNALS

FEATURES
PILOT—MOVES
ROCKETS—MOVE
LIGHT—FLASHES
TRIGGER—CLICKS

SEE BOTTOM OF BOX FOR INTERPLANETARY CODE

IRWIN FLASHING "ROCKET SHIP"
SPACE PISTOL

INSTRUCTIONS
Unscrew nose to
replace battery or bulb.
Use standard bulb or
battery.

111 8x22x15/AUROR/1960'S

Figures 111 through 122: This series of horror characters was made throughout the 1960s, and is still popular now. One of the chief attractions of these toys is that the buyer paints them himself. They are aimed at adults rather than children because they attract fanatics.

112 8x20x25/AURORA/1960'S

113 20x13x23/AURORA/1960'S

114 13.5x9.5x21/AURORA/1960'S

115 22x13x24/AURORA/1960'S

116 11x8.5x22/AURORA/1960's

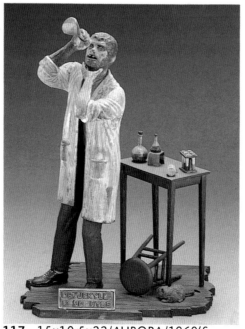

117 15x10.5x22/AURORA/1960'S

118 16x12x22/AURORA/1960'S

119 21x14x15/AURORA/1960'S

120 13.5x9.5x23.5/AURORA/1960'S

121 12.5x12.5x16/AURORA/1960'S

122 16x14x21/AURORA/1960'S

123, 124 19x12x32, 7x5.5x14/MARX/B,W/1950'S: Figure 123 is battery-operated and notable for its extraordinary expression and realistically jerky movements. Figure 124 is a wind-up.

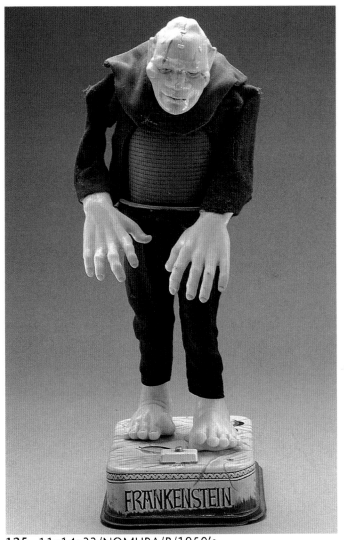

125 11x14x33/NOMURA/B/1950's

126 11x14x33/NOMURA/B/1960'S

127 8X7X17/UNKNOWN/B/1950'S

303

128 13x100x30/MARX/B/1960'S(battery operated): This toy dragon is one meter long. The lights of its tail flash one after another, and it can blow bubbles from its mouth.

129 18x10x28/MARX/B/1950'S

130 21x12x27/MARX/B/1950'S

131,132: 12x15x34/MARX/B/1950'S These wood monsters are very popular in the United States. They can move their eyes and mouths and make a noise like the wind when their leaves are lifted.

133 16x17.5x27/ICHIDA/B/1950'S (battery operated): This is a bank. A fortune appears in the fortune teller's hand every time money is dropped into the bank. Its face also lights up.

Figures 134 through 142: The theme of science fiction appeared in many kinds of toys. Mars and Venus were especially attractive to the children of the 1950s.

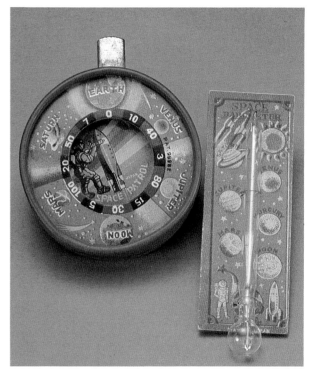

136,137 6.5x6.5x2, 3x9x1/UNKNOWN/1950'S

134,135 2x2x5.5, 4x4x16/UNKNOWN/1950'S

138 8x16.5x9/MASUDAYA/1950'S

139 2.5x14x2.5/UNKNOWN/1950'S

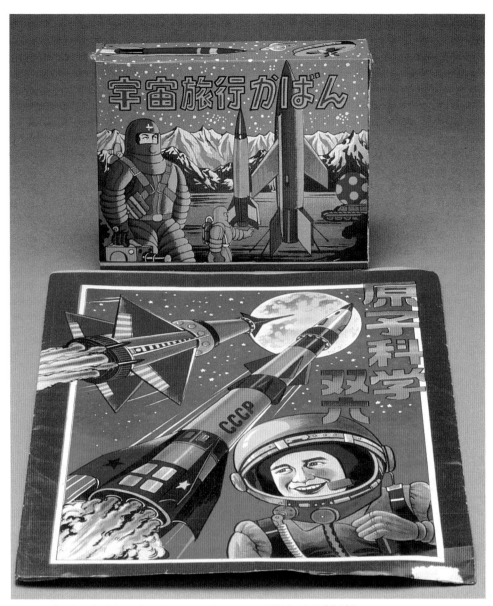

140,141 16x2x12, 22x27/UNKNOWN, MITSUWA/1950'S

142 17.5x4.5x14/SHOYA/1950'S

143 1950'S

144 1950'S

145 1950'S

146 1950'S

147 1950'S

FLIGHT TO MARS

149 1950'S

313

150 1950'S

151 1950'S

152 1950'S

153 1950'S

154 1950'S

地球最后の日

155 1950'S

156 1950'S

157 1950'S

158 1940'S

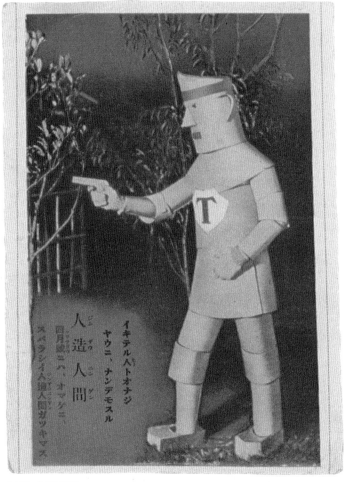

159 1930'S

FUTURISTIC DREAMS

I was brought up in the Kyobashi district of Tokyo in the 1950s. Those were the thrilling years of scientific expansion. It was the decade in which the word *science* came to be a word not just for specialists, but for everyone who felt insatiable curiosity about the future and outer space. Today, science progresses at a remarkable speed; it surpasses the comprehension of the ordinary daydreamer. But we who have been entranced by science have a simpler vision, a vision that was captured in the science fiction toys of our youth. These toys fascinate me now because they embody the unknown of that earlier time—they recall the beauty of the daydreams of my childhood.

—*Teruhisa Kitahara*